T0220270

Notes on the Underground

Notes on the Underground

An Essay on Technology, Society,
and the Imagination

new edition

Rosalind Williams

The MIT Press
Cambridge, Massachusetts
London, England

© 2008 Massachusetts Institute of Technology

All rights reserved. No part of this book may be reproduced in any form by any electronic or mechanical means (including photocopying, recording, or information storage and retrieval) without permission in writing from the publisher.

Set in Bembo by DEKR Corporation.

Library of Congress Cataloging-in-Publication Data

Williams, Rosalind H.
Notes on the underground : an essay on technology, society, and the imagination / Rosalind Williams. — New ed.
p. cm.
Includes bibliographical references and index.
ISBN 978-0-262-73190-4 (pbk. : alk. paper)
1. Underground areas. 2. Underground utility lines. I. Title.
TA712.W55 2008
303.48'3—dc22

2007051623

to my parents

Contents

Preface

When I gave a talk on "imaginary underworlds" in the fall of 1986 at the annual meeting of the Society for the History of Technology (SHOT), I assumed the topic would evolve into a paper of fifty pages or so. In commenting on my talk, however, William Leach remarked in an offhand way that "there is certainly material here for a book." That was the first time the thought of a book-length essay entered my head, and I am much indebted to Bill for suggesting the idea.

A little over a year later, at the annual meeting of the American Historical Association, I presented another talk on the same topic. This time the commentator was Melvin Kranzberg, a founding father and leading light of SHOT. Mel took me to task, in his kind but frank way, for some of the looser ideas rattling around in my summary of what I had come to think of as "the underground book." The resulting manuscript is better for Mel's intervention, though I have no doubt he could suggest ways to make it better yet.

These two episodes underscore what is special about the scholarly environment of SHOT: it offers friendly encouragement and rigorous (though still friendly) criticism, both of which a writer needs. The same blend was offered by Michael Smith (University of California at Davis) and Richard Fox (Reed College), whose detailed and helpful comments on my first draft made the task of revision much easier. I also benefited from the comments of James Paradis, Harriet Ritvo, and Elzbieta Chodakowska. They are all members of my home department, the Writing Program at the Massachusetts

Institute of Technology, and they all took the time to read the manuscript and to offer helpful comments.

Parts of the manuscript were also read by Leo Marx of MIT's Program in Science, Technology, and Society. Leo has given me kindness and guidance ever since he was teaching at Amherst College and I was just embarking upon my doctoral studies. This encouragment over the years has meant more to me than he can imagine.

Most of this book was written during a semester when I was relieved of teaching duties and could work on the manuscript full time. That happy interlude was made possible by a grant provided by the Old Dominion Foundation and administered by MIT's School of Humanities and Social Science, under the leadership of Dean Ann F. Friedlaender. I am indebted to the School for this crucial assistance. I am also grateful to Kenneth R. Manning, head of the Writing Program, for the way he has constantly nurtured my scholarship with both moral and practical support.

In gathering illustrations for this book, I received generous advice and assistance from Merrill Smith and her colleagues at the Rotch Visual Collections at MIT and from Constance Wick at the Bernhard Kummel Library of the Geological Sciences at Harvard. Larry Cohen and Paul Bethge, my editors at The MIT Press, moved the manuscript through the publication process more quickly and painlessly than I believed possible. The experience of working with them has been as enjoyable as it has been educational.

In the old days, a preface usually ended with a male academic thanking his wife for doing the typing and keeping the kids out of his hair. In this less predictable age, I want to thank my husband Gary for helping me with the complexities of word processing and for doing extra babysitting when deadlines pressed. In any age, though, the subtext is the same: the real gratitude is for the confidence that inspires self-confidence and for the love that motivates the practical assistance.

Friends and colleagues too numerous to name, once they discovered what I was writing about, offered suggestions for underground books or topics to include. Many of their ideas were incorporated here—but many others were not, lest the book become far too long and unwieldy. I have repeatedly reminded myself that the purpose of this essay is not to cover the topic but to uncover it.

Readers will inevitably collaborate in this effort as they add their own experiences and ideas. This active contribution—made each time a reader thinks "She should have mentioned . . ."—will be a sign not of the book's failure, but of its success.

Notes on the Underground

The Underground as a Vision of the Technological Future

1

What are the consequences when human beings dwell in an environment that is predominantly built rather than given? This book seeks to answer that question. It explores the psychological, social, and political implications of living in a technological world.

Environment and technology form not a dichotomy but a continuum. The drive to modify the natural or given environment so that it will be safer, more abundant, and more pleasing is as old as humankind. Even landscapes that now seem natural have been profoundly influenced by human activity. Today we admire the olive groves, vineyards, citrus trees, and wheatfields of the Mediterranean basin as emblems of timeless natural beauty, yet none of these crops is indigenous to the area; the terraces, groves, and fields are all products of centuries of laborious human intervention.[1] The human environment has always been, to some degree, artificial.

The degree of the artificiality is what has changed so radically in modern times. The goal of transforming the environment may be ancient, but our ability to realize that goal is unprecedented. In the late twentieth century, our technologies less and less resemble tools—discrete objects that can be considered separately from their surroundings—and more and more resemble systems that are intertwined with natural systems, sometimes on a global scale.

There are objective ways to measure this increasing dominance of the technological environment. Scores of scientific studies have described its effects in quantitative terms: the elimination of animal species (by the year 2000 the number of extinctions is expected to reach 20 percent of all species), the destruction of forest acreage (especially of tropical rain forests), the depletion of groundwater from aquifers (such as the High Plains Aquifer in the United States), the disappearance of chemicals in the atmosphere (most evident in the Antarctic "ozone hole"). Scientists concur that, although human modification of the natural environment is by no means novel, the rate and the extent of the modification are unparalleled.[2] The only historical event that is comparable, in terms of environmental transformation, is the discovery of agriculture.

The physical consequences of this environmental transformation are of supreme importance to human survival. They are not the only consequences, however. This book examines what is commonly called "the environmental crisis" from a cultural rather than a material perspective. The term "environmental quality" is inherently ambiguous: it describes the physical capacity to support life, but it can also refer to an unquantifiable capacity to provide psychic and social well-being. One can imagine a future environment that adequately provides the material basis for human life but is psychologically and socially intolerable. This book asks whether destruction of the natural environment might be culturally as well as physically harmful to human life. The historian of religion Mircea Eliade has reminded us that the Neolithic shift from pastoral to agricultural civilization "provoked upheavals and spiritual breakdowns whose magnitude the modern mind finds it well-nigh impossible to conceive. An ancient world, the world of nomadic hunters, with its religions, its myths, its moral conceptions, was ebbing away. Thousands and thousands of years were to elapse before the final lamentations of the old world died away, forever doomed by the advent of agriculture. One must also suppose that the profound spiritual crisis aroused by man's decision *to call a halt and bind himself to the soil* must have taken many hundreds of years to become completely integrated."[3] It is not only imaginable but probable that humanity's decision to *unbind* itself from the soil—not to return to a nomadic existence, but to bind itself instead to a predominantly technological environment—has provoked a similarly profound spiritual crisis. We are now embarked upon another period of cultural mourning and upheaval, as we look back to a way of life that is ebbing away.

Our sense of mourning is most evident when we see a natural landscape that has been demeaned by technological intrusion: the office park in the meadow, the highway through the marshlands, the fields leveled for malls, the beach walled with condominiums, the canyons sprinkled with tract housing and laced with highways. Such experiences often arouse sadness, apprehension, rage. But these are not our only responses to the technological environment. The same artifacts that are lamented as a violation of nature can, in another context, be celebrated as a victory over nature. We are more likely to admire a technological environment when, instead of seeming to invade and degrade nature, it displaces nature entirely. For example, Houston and St. Paul-Minneapolis have developed elaborate systems of tunnels and walkways that are virtual indoor cities (the Houston system runs 6 miles; the St. Paul skywalks connect 32 city blocks). These surroundings are comfortable both physically and socially, for they exclude not only summer heat and winter cold but also beggars and other people considered undesirable.[4] Shopping centers, airports, and hotels can be charmed worlds of light and color and music. In these artificial environments, the pain of contrast is absent. Here "mechanization takes command,"[5] and every detail of construction is planned to comfort the body and relax the mind. Contemporary experience of the artificial environment is therefore contradictory: while we grieve for a lost way of life, we rejoice in a new one.

How can we come to terms with these conflicting responses? Eliade has proposed that fundamental changes in humanity's control of the material universe also bring fundamental changes in its universe of meaning, which he calls "the imaginary world."[6] For example, he says, when human beings learned to exert control over the material world through mining and metallurgy, they also discovered a new spiritual world. As a student of religion and mythology, Eliade is primarily interested in imaginary worlds as they are expressed in sacred myths and rituals. Secular literature, however, can also embody the world of meaning that evolves along with the material world. If the invention of the pastoral—at once one of the most durable and supple modes of literature—arose from the invention of agriculture and cities, then the great environmental transformation of our own day may lead to other literary inventions equally fundamental and enduring.[7] That remains to be seen. In the meantime, older forms of literature have been used to explore the significance of our new material environment.

The thesis of this book is that, since the nineteenth century, narratives about underground worlds have provided a prophetic view into our environmental future. Subterranean surroundings, whether real or imaginary, furnish a model of an artificial environment from which nature has been effectively banished. Human beings who live underground must use mechanical devices to provide the necessities of life: food, light, even air. Nature provides only space. The underworld setting therefore takes to an extreme the displacement of the natural environment by a technological one. It hypothesizes human life in a manufactured world. What would human personality and society look like then?

The underworld was first defined as an environmental metaphor by Lewis Mumford (b. 1895) in *Technics and Civilization* (1934), his pathbreaking and still irreplaceable study of the machine age in a cultural context. *Technics and Civilization* was published at a pivotal point in Mumford's career. After writing a series of works focusing on American cultural history and literature, he had decided to widen his embrace both in time and space and to attempt a more general interpretation of the triumph of mechanization in the West. To become better acquainted with European sources in technological history, Mumford applied for and received a Guggenheim fellowship, which enabled him to take a four-month trip to Europe in early 1932. On that trip he visited Munich's Deutsches Museum, where he was especially captivated by the novel and realistic life-size reproductions of ore, salt, and coal mines.[8]

In writing the early chapters of *Technics and Civilization,* Mumford drew upon this impressive experience to stress the historical significance of mining in promoting early industrialization (the steam-and-coal phase, which he called "paleotechnic industry"). Mumford went on, however, to explore the mine's metaphorical significance as a working environment. Whereas other environments contain some sort of food, some product directly translatable into life, Mumford noted, "the miner's environment alone is—salt and saccharin aside—not only completely inorganic but completely inedible."[9] The peasant on the surface of the earth may also endure backbreaking labor, but he has the beauties of nature to distract him: the scurrying rabbit that recalls the pleasure of hunting, the passing girl who reminds him of his manhood, the play of light on the river that awakens reverie. The miner, however, works in a place from which organic nature seems to have been banished:

The mine . . . is the first completely inorganic environment to be created and lived in by man: far more inorganic than the giant city that Spengler has used as a symbol of the last stages of mechanical desiccation. Field and forest and stream and ocean are the environment of life: the mine is the environment alone of ores, minerals, metals. . . . Except for the crystalline formations, the face of the mine is shapeless: no friendly trees and beasts and clouds greet the eye. . . . If the miner sees shapes on the walls of his cavern, as the candle flickers, they are only the monstrous distortions of his pick or his arm: shapes of fear. Day has been abolished and the rhythm of nature broken: continuous day-and-night production first came into existence here. The miner must work by artificial light even though the sun be shining outside; still further down in the seams, he must work by artificial ventilation, too: a triumph of the "manufactured environment."[10]

In earlier drafts of this section of *Technics,* Mumford experimented with other language to express his perception of the mine as an environmental model. In his first draft of 1933, Mumford described the mine as an "environment of work, unremitting and undistracted work." He kept this idea (in *Technics* itself the sentence reads "Here is the environment of work: dogged, unremitting, concentrated work"[11]), but he knew it was incomplete: in the margin he scrawled a note to himself: "Necessity to live by artificial means."[12] On his next draft Mumford added a version of the passage quoted above, noting that mining requires artificial light and ventilation. In this second draft, however, the passage ends with the words "a triumph, too, of the non-environment."[13] By the final draft, Mumford realized that the term "non-environment" was not quite the right term either, since life, by definition, takes place in some environment. The metaphorical significance of the mine is not the absence of environment but the dominance there of "artificial means." This, finally, is how Mumford defines mining as a metaphor for modern technology in *Technics:* as a place where the organic is displaced by the inorganic, where the environment is deliberately manufactured by human beings rather than spontaneously created by nonhuman processes.

When Mumford put "manufactured" in quotation marks in the final draft, he indicated his awareness that he was inventing a term, that no standard term was available—which is still the case.[14] There was, however, a standard *image* of the manufactured environment: the city. As Mumford noted in *Technics,* Oswald Spengler had used the giant city as a symbol of "the last stages of mechanical desiccation" in his enormously influential work *The Decline of the West* (1928). According to Spengler, as the soul of the city develops, it becomes the whole world; the gigantic megalopolis "suffers nothing beside itself and sets about *annihilating* the country picture." There is no question of coexistence with nature, which is banished: ". . . here the picture is of deep, long gorges between high, stony houses filled with coloured dust and strange uproar, and men dwell in these houses, the likes of which no nature-being has ever conceived. Costumes, even faces, are adjusted to a background of stone. By day there is a street traffic of strange colours and tones, and by night a new light that outshines the moon."[15] In much the same spirit (for he borrowed heavily from Spengler's interpretation), Mumford foresees in *The Culture of Cities* (1938), his next book after *Technics,* the self-destructive development of an independent urban environment. According to Mumford, the already dominant Metropolis will become a devouring Megalopolis and finally a wasted Necropolis. Even now, city streets are coming to resemble deep pits: "Nature, except in a surviving landscape park, is scarcely to be found near the metropolis; if at all, one must look overhead, at the clouds, the sun, the moon, when they appear through the jutting towers and building blocks."[16]

Countless other writers, from Edward Bellamy (*Looking Backward,* 1888) to Yevgeny Zamyatin (*We,* 1920) to Aldous Huxley (*Brave New World,* 1932),[17] have used the city as an image of an artificial environment. The imagery of hyperurbanity resembles that of the underground. In this century, imaginary underworlds are almost always vast cities and many "fantastic cities" are buried far below the ground.[18] Even cities on the surface are imagined as being so detached from nature that they resemble caves.[19] Spengler had predicted that the giant cities of the future would be stone masses harboring a decadent population of neo-cavedwellers.[20]

In *Technics* Mumford himself had written that nineteenth-century cities were, in fact and in appearance, extensions of the coal mine.[21] Mumford insisted, however, the underworld is a more effective image of a manufactured environment, in that it is "far more inorganic" than the city. He was keenly aware that, until fairly recent times, even cities as large as London or New York or Paris had retained a rural character. As a boy growing up in New York City around the turn of the century, Mumford had daily strolled around gardens, trees, animals, ponds, and rivers; "the Port of New York," he wrote in his autobiography, "became my Walden Pond."[22]

In the decades after writing *Technics and Civilization* and *The Culture of Cities,* Mumford continued to search for images to convey the all-encompassing nature of a technological environment. He struggled to find a vocabulary that would adequately express his conviction that the twentieth century was witnessing an unprecedented stage in technological development. In his research and writing, Mumford kept pushing the origins of mechanization back to earlier and earlier times—back to ancient Egypt, and eventually to the discovery of agriculture. As he had earlier enlarged his compass from American to Western history, now he embraced the globe. From this sweeping perspective, Mumford concluded that humanity was moving into an unprecedented stage of mechanization, one motivated by a new technological ideal. In the 1960s he invented the term "megatechnics" to describe that ideal: "In terms of the currently accepted picture of the relation of man to technics, our age is passing from the primeval state of man, marked by the invention of tools and weapons, to a radically different condition, in which he will not only have conquered nature but detached himself completely from the organic habitat. With this new megatechnics, he will create a uniform, all-enveloping structure, designed for automatic operation."[23]

Today, not the city but the spaceship has become the standard image of the megatechnic ideal of complete detachment from the organic habitat. This newer image, however, denies the claustrophobic realities of human life on earth. Although the spacecraft itself may model an all-encompassing technological environment, its mission—hurtling through endless space, going boldly where no one has gone before—suggests the vision of an endless frontier where new varieties of nature wait to be discovered. Unlike the mine, the spaceship fails to convey a sense of permanent enclosure in a finite world. Furthermore, because of the indeterminacy of the interstellar void, space travel lacks the verticality that gives the underworld its unique power in the human imagination.

Stories of descent into the underworld are so ancient and universal that their fundamental structure, the opposition of surface and depth, may well be rooted in the structure of the human brain. The congruence may be explained by the Freudian hypothesis of an Oedipal experience that splits human beings into conscious and unconscious selves, or by the Jungian hypothesis of a collective subconscious. In any case, the metaphor of depth is a primary category of human thought.

It is the combination of enclosure and verticality—a combination not found either in cities or in spaceships—that gives the image of an underworld its unique power as a model of a technological environment. If we imagine going underground, we not only imagine an environment where organic nature is largely absent; we also retrace a journey that is one of the most enduring and powerful cultural traditions of humankind, a metaphorical journey of discovery through descent below the surface. The primary documents for this study will be nineteenth-century fictional narratives that imagine human life in subterranean space. Before looking at these narratives, though, we need to look at the cultural tradition from which they emerge.

Long before Virgil's Aeneas was guided by a Sibyl to the infernal regions through a cave on the leaden Lake Avernus, long before stories of Proserpine's abduction to the underworld by Pluto or of Orpheus's descent to the Stygian realm to bring back Eurydice,[24] and long before recorded history, when the earliest humans drew the bison and bears they hunted on the walls and ceilings of caves, they must have told stories about the dark underworld lying even deeper within the earth. Even in environments that lack caves—the Kalahari Desert, and the flat open landscapes of Siberia and Central Asia—the preliterate inhabitants assumed a vertical cosmos: sky, earth, and underworld. The underworld might be a region of water, or fire, or a counterheaven (suggested by the way the sun and stars dip below the horizon), but in any case nature was assumed to be as deep as it was high, its major axis vertical.[25] In this richly symbolic universe of the past, vertical movement was far more significant than movement in the horizontal axis. Narratives about journeys to the world below were inherently sacred.

The idea of a vertical cosmos began to weaken in Europe during the age of the great explorations, between about 1500 and 1700.[26] At the same time, as part of the same gradual process of secularization, the sacred myths that form the earliest content of Western literature began to be displaced by fictive forms that "undermined the religious authority of those [mythic] structures by constituting a kind of secular substitute for them."[27] Beginning in the Renaissance, the epic tradition of the journey to the underworld was transformed into narratives that were written and secular rather than oral and sacred. In these narratives, an adventurous, unlucky, or half-mad traveler sets forth and discovers an underworld, which he enters and from which he may or may not emerge.

In some of these narratives, the imprint of the earlier sacred tradition is still quite evident. For example, William Beckford's (1760–1844) widely read *Vathek* (1787) tells how the Caliph Vathek, a man of prodigious powers and appetites, enters into a pact with Eblis, the Oriental Satan. After renouncing his religion and God, the caliph is allowed to enter Eblis's Palace of Subterranean Fire, which lies below the ruins of an ancient city and which holds treasures and talismans. When Vathek and his lover approach the ruined city, a rock platform opens before them and a polished marble staircase leads them downward to the realm of Eblis: ". . . they found themselves in a place which, although vaulted, was so spacious and lofty that at first they took it for a great plain. As at length their eyes became accustomed to the great size of surrounding objects, they discovered rows of columns and arcades running off in diminishing perspective until they concentrated in a radiant spot like the setting sun painting the sea with his last rays."[28] Then they see an immense hall holding a multitude of pale specters, some shrieking, others silent, all with glimmering eyes and with their right hands on their hearts, which are consumed by fire. This is not heaven but hell; Vathek and his lover too begin to burn with hatred and are condemned to eternal despair. *Vathek* is a self-conscious, precious, but compelling revival of the sacred tradition by a highly sophisticated writer (it was composed in French by a wealthy English aristocrat-aesthete). It powerfully influenced Byron and Keats, and it has been praised by Jorge Luis Borges as "the first truly atrocious Hell in literature."[29]

Many other underworld narratives are considerably more lighthearted. A generation before *Vathek* a wandering Norwegian scholar, Baron Ludvig Holberg (1684–1754), wrote what became the best-known early subterranean adventure tale, *The Journey of Niels Klim* to the *World Underground* (1741). Holberg's hero Klim is exploring a cave when his rope breaks. He ends up (or rather down) in the underground land of Potu, which he discovers is inhabited by rational, peripatetic trees. On further exploration Klim finds that Potu is one kingdom on another planet, Nazar, which lies hidden inside the planet Earth. Nazar too has a sun that rises and sets, although night is not very dark because light from the subterranean sun is reflected from Earth's inner surface.

The Journey of Niels Klim is an "imaginary voyage," a type of narrative that emerged in the late 1400s, when actual voyages of discovery were reshaping Western civilization. In these narratives the description of the supposedly discovered land or lands, besides being entertaining, is often intended to comment upon contemporary society by presenting an alternative at once strange and recognizable. The best-known imaginary voyage in English is Jonathan Swift's *Gulliver's Travels,* but there are many others; one scholar has cataloged 215 such tales written between 1700 and 1800.[30] The plot devices are simple and oft-repeated: a shipwreck on a mysterious island (used by Shakespeare in *The Tempest*), a new invention (Johannes Kepler used this in a tale about a voyage to the moon)—or the discovery of a subterranean world.[31]

This narrative tradition retained its popularity through the nineteenth century. Jules Verne titled his series of novelistic adventures "imaginary voyages," and H. G. Wells used the time-honored device of a seemingly magical invention (a time machine, a gravity-defying substance) to carry explorers into underground worlds distant either in time (far into the earth's future) or in space (inside the moon). In the nineteenth century, however, another type of underground story also began to be written—one quite different in its fundamental structure. Instead of being a place to visit, the underworld becomes a place to live. Instead of being discovered through chance, an underworld is constructed (or a natural underworld is vastly enlarged) through deliberate choice.

The difference between the two types of stories is evident if we compare two works by Jules Verne. In 1864 Verne published *Voyage au centre de la terre* [Journey to the Center of the Earth], the story of the Danish professor Lidenbrock and his nephew Axel, who, along with a guide, descend into the crater of an Icelandic volcano, explore a maze of caverns, discover an underground sea, glimpse a giant man-like creature driving a herd of mammoths, and finally burst back onto the earth's surface through Mount Etna. Then and now, *Journey* is one of Verne's most popular books. Far less famous, and quite different in its premise, is another Verne book on a subterranean theme, published thirteen years later. *Les Indes noires* (1877) [variously translated as The Black Indies, Black Diamonds, Underground City, or The Child of the Cavern] is a story not of intrepid explorers and guides but of hard-working engineers and miners, and not of an exciting journey but of a permanently functioning underground society. The story goes that a Scottish miner, unwilling to believe that the pit he worked had been exhausted, moved with his family into the mine after it was closed. There he doggedly searched for a new seam. Eventually he found one, and other miners join him to build a subterranean, utopian Coal Town—an autonomous underworld environment, artificially lighted and ventilated and capable of supporting human life in complete independence from the surface world. One novel describes an imaginary subterranean journey, the other an imaginary subterranean society.

To explain the emergence of the second type of underground tale, we must look outside literary tradition. The underground may be an enduring archetype, but it is not a theme beyond time and society. The idea of permanently living below the surface of the earth emerged along with modern science and technology. As scientific knowledge advanced, the idea of discovering a hidden inner world became less and less credible. As technology advanced, on the other hand, the idea of building an inner world became more and more credible. Let us look more carefully at each of these developments.

At the time *Niels Klim* was written, in the middle of the eighteenth century, the possibility that the earth was hollow and habitable within still had some respectable advocates. In the early 1700s the French civil engineer Henri Gautier (1660–1737) combined some Cartesian theories of gravity and some experiments with a mercury barometer to conclude that at 1,195 fathoms below sea level gravity would go to zero and then become negative, and that 1,195 fathoms still lower one would find an inner sea, the mirror image of the external one. The two oceans, Gautier theorized, were connected at the poles, and the earth was as hollow and light as a balloon.

In the late 1700s, however, other geologists (the term *geology* was beginning to be used about that time), using not Cartesian but Newtonian physics, convincingly demonstrated that the earth as a whole, far from being hollow, must be far denser than its outlying rocks.[32] In 1774 Nevil Maskelyne, the Astronomer Royal, theorizing that the presence of mountains would affect a plumb line as Newton had predicted in 1729, realized that if he could quantify this effect he could calculate the average density of the earth. He set up an observatory on either side of a ridge in northern Perthshire and determined the direction of the plumb line with respect to selected stars at each of the two locations. According to Maskelyne's calculations, the mean density of the earth was double that of the hill. (His estimate was high; the underlying rocks in the area are quartzites, marbles, and mica schists, which give it a higher-than-average density.) In 1798, Henry Cavendish (1731–1810) carried out experiments with an elegant torsion balance that measured the Newtonian gravitational attraction between bodies of known masses, thus permitting him to "weigh the earth." He calculated that the average density of the earth is 5.48 times that of water. Cavendish's work was more accurate (the currently accepted value is 5.517), but Maskelyne's work, which was made on a real mountain, was more impressive to many geologists. In reporting his work, Maskelyne scoffed at Gautier and others "who supposed the earth to be only a hollow shell of matter."[33]

Despite such evidence of the earth's density, the idea of a hidden inner world persisted into the nineteenth century.[34] John Cleves Symmes (1780–1829), an eccentric officer in the U.S. infantry, theorized that the earth consists of five hollow concentric spheres, with spaces between them, and that these spheres are habitable on both their convex and their concave surfaces. The inner realm could be entered through polar openings thousands of miles in diameter. Around each opening there was an icy "hoop," but within the hoop the climate was mild or even hot. Despite evidence of the earth's density, this theory received considerable attention when in 1823 Symmes petitioned Congress to sponsor an Antarctic expedition to search for the polar opening. Symmes's disciple Jeremiah Reynolds (1799–1858) carried on after Symmes's death, stressing the general benefits that could be derived from such an expedition: the potential contribution to the whaling industry, the charts that would aid naval and merchant ships, the strengthening of trading ties with Latin

America. Thanks in part to Reynolds's efforts, an expedition to chart the Antarctic coast was funded by Congress and set sail in 1838. It consisted of six vessels carrying, along with a crew of "invalids and idlers" (in words of its captain, Charles Wilkes), seven civilian scientists, two artists, and two technicians—the largest group of civilian researchers assembled in federal service at that time. The observing, the collecting, and the mapping done on the 85,000-mile Wilkes Expedition make it a milestone in the development of American science. In other respects, the expedition proved calamitous. Only 181 of the original crew of 346 men returned in 1842; the rest were lost to disease, shipwrecks, or desertion.[35]

Symmes's lasting contribution was not to scientific fact but to literary imagination. Edgar Allan Poe reviewed Reynolds's 1834 speech to the House of Representatives in *The Southern Literary Messenger* and later defended him in other publications.[36] Poe was entranced by the possibility of a subterranean realm; visions of such a world appear in his works from "MS. Found in a Bottle" (1833) to "Eureka" (1848), including *The Narrative of Arthur Gordon Pym,* "Hans Pfaal," "Dream-Land," and "Ulalume." In *Arthur Gordon Pym,* the shipwrecked hero floats southward on a raft in an iceless sea, under skies lit by strange flares of aurora borealis, until he finally rushes into a gigantic cataract—evidently the polar opening where the outer ocean pours into the inner one—and glimpses there a huge, shrouded, perfectly white human figure. In this tale Poe quotes verbatim almost half of Reynolds's 1,500-word 1834 appeal to Congress. An even more powerful stimulus to Poe's imagination appears to have been Symmes's novel *Symzonia: A Voyage of Discovery,* published in 1820 under the pseudonym "Captain Adam Seaborn."[37] It is a lively, often unintentionally amusing account of a polar expedition that comes upon a mysterious ship of unearthly swiftness ("no drift from the external world," intones Captain Seaborn) and finally sails into the Internal World, a verdant and rolling land gently lit by the reflected light of the sun entering through the polar opening and inhabited by a wise and rational race of "Internals." From this entertaining but strange utopian tale,[38] Poe fashioned some of his most haunting images: the storm and the mystery ship that appear in "MS. Found in a Bottle," the land of spirits inside the earth discovered by Pym, the polar opening seen by Hans Pfaal on his way to the moon, the mysteriously lighted images of "Ulalume" and "Dream-Land." In the hours before his death in the charity ward of a Baltimore hospital, raving in a night-long delirium, Poe repeatedly called out "Reynolds! Reynolds! Oh, Reynolds!"

Jules Verne ardently admired Poe and was well acquainted with *The Narrative of Arthur Gordon Pym.* In 1863 Verne wrote an essay on Poe in which he analyzed the Pym story at length, quoting the last entry in Pym's journal and commenting: "And so the story ends, unfinished. Who will ever complete it? A bolder man than I, and one more bent on venturing into the domain of the impossible."[39] When Verne wrote this essay, he had just published his first great success, *Cinq Semaines en Ballon* [Five Weeks in a Balloon], and he was beginning work on a new novel, this one about the discovery of a polar volcano by a British expedition to the Arctic. Verne had heard of the Wilkes Expedition, and he was intrigued by the idea that active volcanoes at the poles caused openings there and perhaps even provided entrances into the earth's interior. In the new novel (eventually published in 1865 under the title *Voyages et aventures de Captaine Hatteras*), one of the characters, Dr. Clawbonny, firmly supports the idea that the polar regions are habitable, and remarks: "In recent times it has even been suggested that there are great chasms at the Poles; it is through these that there emerges the light which forms the Aurora, and you can get down through them into the interior of the earth."[40]

Verne clearly considered Dr. Clawbonny's theories farfetched, but he was also intrigued by the somewhat more credible speculation that there might be other reasons for hollow spaces under the earth's surface. Geologists were still vague about the way temperatures and densities were distributed inside the planet. The composition of the inner earth was still a mystery, and could be described and interpreted in a variety of ways. According to the eminent geologist Archibald Geikie, who wrote the article on geology in the ninth (1887) edition of the *Encyclopaedia Britannica,* various theories ("mostly fanciful") about the inside of the earth had been propounded, but only three merited serious consideration: the earth might have a solid crust and molten interior; a liquid substratum might lie beneath the crust, the rest of the globe being solid; or the planet might be solid and rigid to the center, except for "local vesicular spaces." The last theory was favored by the eminent Lord Kelvin, co-discoverer of the second law of thermodynamics.[41]

During the winter of 1863–64 Verne had a series of conversations with Charles Sainte-Claire Deville, a geographer who had explored European volcanoes and who theorized that they might be connected by passages under the earth. Verne became so excited by Deville's ideas that he dropped his work on *Captaine Hatteras* and quickly wrote *Journey to the Center of the Earth.* In that book he described the underworld in a way compatible with Kelvin's theory of "local vesicular spaces." Furthermore, since geologists of his day agreed that the earth's interior must be hot (citing as proof the eruption of volcanoes and the thermal gradient in mine shafts),[42] Verne took care to explain how his explorers could descend so far without burning up. *Journey* is farfetched, but not too much so.

Over the next three decades Verne continued to write his memorable "imaginary voyages," including several more on subterranean themes. Eventually, in 1897, Verne summoned the will to take up the challenge he had raised years before in his essay on Poe: completing the Pym narrative, and, in particular, explaining the great white figure that suddenly looms before Pym at the end of the tale. In his 1897 work *Le Sphinx des Glaces* [The Ice Sphinx], Verne returned to the idea of a polar mountain, which had inspired *Captaine Hatteras;* this time, however, the mountain is an enormous sphinx-like lodestone that attracts all the iron in the area. The hero, Jeorling, disparages Arthur Gordon Pym's story as "unadulterated fantasy—delirious into the bargain"—until, at the end of the book, he and his companions discover Pym's corpse on the rock, hanging from the strap of his rifle, which had been irresistibly drawn there by the magnetic sphinx.[43]

By a remarkable coincidence, while Verne, aging and ill, was laboring on *Le Sphinx,* another writer of science fantasy—a vigorous young man not quite thirty and just beginning his career—published an underground tale also featuring a white sphinx. The year was 1895, the writer was H. G. Wells, and the story was *The Time Machine.* In Wells's novel the sphinx-like object turns out to be a hollow, bronze-plated pedestal into which the narrator's time-travel machine is snatched by the sinister underground Morlocks, who, in the world of the year 802,701, hunt the hapless above-ground Eloi. Wells was hailed as another Jules Verne. He was also frequently compared to Edward Bulwer-Lytton, since *The Time Machine* reminded English readers of that author's 1862 book *The Coming*

Race.[44] Within the next six years, Wells wrote several more tales about subterranean realms of technology: *When the Sleeper Wakes* (1899),[45] *A Story of the Days to Come* (1899), and *The First Men in the Moon* (1901).

In the opening years of the twentieth century, the skein of underground stories became even more tangled as authors continued to respond to one another's subterranean visions. In 1905 Wells wrote a long and laudatory foreword to *Underground Man,* the English translation of a utopian fantasy by the eminent French sociologist Gabriel Tarde. (The French version, titled *Fragment d'histoire future,* had been published in 1896.) Once it appeared in English, Tarde's tale was constantly compared with Wells's.[46] A few years later, in 1909, Wells's compatriot E. M. Forster portrayed a dystopian underground society in his short story "The Machine Stops." Forster later explained that "The Machine Stops is a reaction to one of the earlier heavens of H. G. Wells."[47] The "heaven" Forster had in mind was probably Wells's *A Modern Utopia* (1904)—not the earlier *Time Machine,* which is so similar to "The Machine Stops" that one critic has remarked that they "are basically the same story."[48] All these are stories not of descent into a sacred timeless realm, but of projection into a highly technological human future.

Text and context interpenetrate. Events in science and in technology altered the lines of the underground stories, opening up new imaginative possibilities while closing down others. On the other hand, the scientific and technological events were informed by the storyline of the journey to the underworld in quest of truth and power. The significant relationship between literature and science and technology is to be found in this structural congruence of plotlines, rather than in the all-encompassing fog of a zeitgeist or in the association of particular literary images with particular biographical details. At least, this is the argument I shall try to make in the next two chapters.

Chapter 2 addresses scientific developments of the late eighteenth century and the nineteenth century; chapter 3 covers technological developments of the same period. One purpose of these chapters is simply to show how much actual excavation—from the digging up of ancient Troy to the digging of railway tunnels through the Alps—was going on in the period when Verne, Wells, and Forster

---

were writing. They could not avoid seeing some excavation projects and hearing about many others. Even more important, they knew the same was true of their readers. This assumption of shared experience allowed Wells to have the narrator of *The Time Machine* explain to his readers of the 1890s that, although the evolution of an underground species might seem grotesque, "even now there are existing circumstances to point that way":

There is a tendency to utilise underground space for the less ornamental purposes of civilisation; there is the Metropolitan Railway in London, for instance, there are new electric railways, there are subways, there are underground workrooms and restaurants, and they increase and multiply. Evidently, I thought, this tendency has increased till Industry had gradually lost its birthright in the sky. I mean that it had gone deeper and deeper into larger and ever larger underground factories, spending a still-increasing amount of its time therein, till, in the end—! Even now, does not an East-End worker live in such artificial conditions as practically to be cut off from the natural surface of the earth?[49]

But context involves more than common intellectual and technological experiences. If the mythological journey to the underworld shaped the nineteenth century's subterranean narratives, it also shaped that century's science and technology. The quest to recover the truth about the past by digging ever more deeply was a central project of nineteenth-century science. In technological projects too, excavation was cast in mythological terms, as a heroic journey into forbidden realms. The second and third chapters will study the subterranean quest as a plotline in nineteenth-century science and technology.

Beginning with chapter 4, the emphasis will begin to shift from actual to metaphorical excavation, and from secondary to primary literature. The pivotal point is aesthetic sensibility. New aesthetic concepts—first sublimity and later fantasy—were invented that expressed the emotional power of subterranean environments, a power not encompassed by the traditional aesthetic terminology of beauty and ugliness. Furthermore, sublime and fantastic images were extended from subterranean environments to technology in general. Sublime images dominated the first industrial revolution, while fantastic ones characterized the second industrial revolution. Both aesthetic categories helped shape and organize the sense impressions of daily life—concrete, vivid, but chaotic—into an imaginary world of human meaning. Iconography therefore provides a link between actual excavation and the literary theme of underworld life.

The literature is analyzed in more detail in chapters 5–7. These last three chapters describe and evaluate significant late-nineteenth-century and early-twentieth-century stories about imaginary underground societies. The fifth chapter centers on the social implications of subterranean life, especially the concern that humans living in an enclosed environment might degenerate into feeble, brainless hedonists or cruel, heartless barbarians. Chapter 6 explores the political implications of living in an environment that requires a high degree of planning and in which resources are strictly limited. The note of anxiety evident in chapters 5 and 6 becomes even more pronounced in the final chapter, where the potential for catastrophe in subterranean life is considered. In this last chapter a recurring nightmare is the threat of ecological disaster, caused either by human folly or by nature's revenge.

I originally intended to deal with American as well as with British and French literature. (In the latter case, I am limiting myself to the European language and literature I know best.) With respect to scientific, technological, and aesthetic background, it seemed to me that similarities in experience would outweigh any national differences separating the Americans from the British and the French. I expected to use subterranean stories by American writers such as L. Frank Baum, who wrote numerous sequels to *The Wonderful Wizard of Oz* (1900) set in the underground realm of the Nome King,[50] and Edgar Rice Burroughs, who followed *Tarzan of the Apes* (1912) with a series of tales set in a weird inner world called Pellucidar, inhabited by creatures like the Horibs—snake-men with scales and soft white bellies.[51] As I continued to read and write, however, I discovered that the American stories, while entertaining and significant in their own way, did not fit together well with those written by British and French authors. Their themes did not seem to mesh.

I have concluded that this is no accident. My work supports Leo Marx's contention that American fables expressing the pastoral ideal are cultural symbols arising from the unique American historical experience.[52] As Marx has shown, American writers typically develop the theme of the technological environment on the horizontal plane: the machine invading the garden, the individual retreating, at least temporarily, from complex civilization to a more natural environment out west or in the woods. In particular, American writers explore a "middle landscape," a pastoral haven lying (in their imaginations) between urban civilization and primitive nature. Marx theorizes that this literary tradition is inextricably related to the unique environmental conditions of American national experience: a technically advanced society inhabiting a continent rich in natural resources and sparse in population.

In a similar way, the fable of the technological underworld seems to resonate with Old World conditions, where primitive nature was largely absent and where the presence of the built environment was far more dominant. British and French writers seem to favor a vertical axis that is more uncompromising than the horizontal one prevalent in America. Instead of three zones, they work with only two: surface and subsurface, pastoral and manufactured. The choices are starker because there is no longer a middle landscape; primitive, raw nature is no longer perceived as an alternative. Furthermore, vertical movement between subsurface and surface is far more difficult than the relatively easy horizontal passage between city and pastoral retreat.

As these comments illustrate, by their very nature these stories practically demand a structuralist reading that focuses on the recurrent spatial patterns. Like the myths to which they are so closely related, subterranean narratives can be analyzed into repetitive units—for example, the opposites of surface and depth—which combine and recombine like elements of grammar.[53] Certainly my reading of these stories owes much to techniques of narrative analysis that, in turn, owe much to structuralism. But if we begin here, we cannot end here. If we focus too exclusively on schematic repetitions, the stories all begin to look the same.[54]

In fact, they are not at all the same, because other elements of narrative keep escaping from the relatively rigid common structure. Narration permits and indeed encourages contradiction, exploration, questioning, and suspension of judgment, as opposed to abstract and logical statements of conviction. The story can move in directions that the writer did not foresee at the outset. Narratives create their own momentum because the very act of storytelling opens up unexpected and perhaps unintended possibilities. In this respect the narrative process is like the technological one: the act of construction develops its own momentum.[55]

Even more obviously, the exploratory quality of narrative method resembles that of scientific inquiry (although one should not push the comparison too far; there are many important distinctions between fictional narrative and scientific theorizing). The common point is the opportunity—indeed the need—to expand from the present to the future.[56] H. G. Wells called *When the Sleeper Wakes* one of his "fantasias of possibility." These fantasias, he said, "take some great creative tendency, or group of tendencies, and develop its possibility in the future." "'Suppose these forces to go on,'" said Wells, "that is the fundamental hypothesis of the story."[57] Scientists use a similar process when they run a computer model to test a hypothesis; they calculate, on the basis of certain assumptions, what will happen if a "group of tendencies" should go on in the direction in which it is now headed. In many cases (especially when dealing with nonreproducible, large-scale environmental events) they must project possibilities rather than confirmed data. This mode of thinking has been described by the anthropologist Clifford Geertz as "neither more nor less than constructing an image of the environment, running the model faster than the environment, and then predicting that the environment will behave as the model does."[58]

The defining characteristic of the subterranean environment is the exclusion of nature—of biological diversity, of seasons, of plants, of the sun and the stars. The subterranean laboratory takes to an extreme the ecological simplification of modern cities, where it sometimes seems that humans, rats, insects, and microbes are the only remaining forms of wildlife. The guiding principle of this literary experiment is, therefore, what the critic Fredric Jameson has termed "world-reduction": "a principle of systematic exclusion, a kind of surgical excision of empirical reality, something like a process of ontological attenuation in which the sheer teeming multiplicity of what exists, of what we call reality, is deliberately thinned and weeded out through an operation of radical abstraction and simplification."[59] Jameson reminds us that the "high literature" of the nineteenth century explicitly sought a "cognitive and experimental function," whether through "world-reduction" or through other principles.

That "cognitive and experimental function," though, is practiced not in the natural world but in the social world—and here the analogy between literature and science breaks down. Over and over again, as we shall see, what is most simplified in these narratives is not nature but humanity. The writer descends below the social surface to seek the truth about the lower classes—people regarded primarily as subjects for cognition and research, as pieces of buried evidence to be dug up and analyzed. What the writer unearths, however, is less fruitfully read as a description of lower-class reality than as a reflection of middle-class anxieties that the writer shares.[60] These stories have to be read not only as mythic structures, not only as prophetic narratives, but also as expressions of bourgeois consciousness. Only then do we become aware of the extent to which nature and technology are both class-related categories. The technological environment is regarded as a threat primarily because of the social arrangements it seems to imply; conversely, nature is presented as an alternative not so much to an undesirable technological order as to an undesirable social order.

At the outset I said that the goal of this book is to explore the psychological, social, and political implications of living in a predominantly technological environment. In order to deal with all these consequences, we must read these subterranean stories simultaneously as myths, narratives, and ideologies, without letting any one of those readings overpower the rest. That is quite a challenge, but only then can we assess the full significance of the present environmental transformation. The subterranean environment is a technological one—but it is also a mental landscape, a social terrain, and an ideological map.

Excavations I: Digging Down to the Truth

2

In *Technics and Civilization,* while discussing the mine as a model of the inorganic environment, Lewis Mumford suggests that it also serves as a model for the modern scientific worldview: "It is a dark, a colorless, a tasteless, a perfumeless, as well as a shapeless world: the leaden landscape of a perpetual winter. The masses and lumps of the ore itself, matter in its least organized form, complete the picture. *The mine is nothing less in fact than the concrete model of the conceptual world which was built up by the physicists of the seventeenth century. . . .*"[1] The questions Mumford raises here touch upon a mighty theme: the disenchantment of the world (Max Weber's famous phrase), or, even bolder, the death of nature (the title of a book by Carolyn Merchant). According to their common analysis, once nature is subjected to scientific rationalism it ceases to be a vital source of human meaning and becomes a matter of fact rather than a matter of value. Modern science views nature as colorless, shapeless, and devoid of any symbolic aura or spiritual significance—not a living world, but a dead mine.

This chapter questions that analysis. Historians of science have long recognized that the origins of modern science are mingled with magical and religious modes of thought. They tend to assume, however, that these modes were gradually shed in favor of rationalistic, quantifying, secular ones. I will argue here that the older, value-laden modes of thought persisted through the nineteenth century. In the words of one mid-nineteenth-century commentator, "a remnant of the mythical lurks in the very sanctuary of science."[2] The historical sciences of that century—geology, paleontology, anthropology, and archaeology—were repeatedly constructed as a mythological narrative. Their common plotline is a descent into the underworld in quest of truth. The imaginative power of that plot explains, in large measure, the profound cultural impact of the nineteenth-century historical sciences.

In these scientific quests, the truth that was sought underground was the answer to the mystery of lost time. The stratified cosmos of the prescientific age was associated with a cyclical concept of time—a concept "modeled on the recurrent phases of nature," the endless cycle of seasons or wheeling of the stars.[3] The historical sciences gave a vertical dimension to time; they discovered "deep time." Geology extended our temporal yardstick from thousands of years to millions and then to billions, until we were able to discern (in the famous words of James Hutton, usually credited as a primary discoverer of geological time) "no vestige of a beginning,—no prospect of an end."[4] Stephen Jay Gould has called the discovery of deep time "geology's greatest contribution to human thought," a discovery "so central, so sweet, and so provocative, that we cannot hope to match its importance again."[5]

This chapter will examine the origins of the idea implied by the phrase "deep time": that time is correlated with space, that digging down into the earth is also going back into the past. Certainly Gould is correct in denouncing the "empiricist myth" that attributes a mystique to fieldwork and that correspondingly underestimates the importance of theoretical preconceptions.[6] He argues persuasively that concepts of time (particularly Hutton's concept of cyclical time) led to fieldwork rather than resulted from it. In cultural terms, though, the significant point is precisely that the activity of fieldwork came to assume mythological status. Whatever its objective importance in the advancement of scientific understanding, the image of excavation took on immense cultural importance. Excavation was seen as a modern version of the mythological quest to find truth in the hidden regions of the underworld. As a result, excavation became a central metaphor for intellectual inquiry in the modern age.

What follows, then, is not a summary of geological and archaeological science in general but a summary of excavation as an ever more ambitious, ever more precise methodology. This opening up of the earth as a subject for scientific investigation altered but did not destroy the mythical dimension of the underground. The earth's inner space may no longer be regarded as sacred, but it still is a repository of spiritual value because it is assumed to hold the secrets of lost time. In this archive is imprinted the story of the origins of man, of the globe, even of the galaxy. The symbolic significance of the underground has changed decisively, but it has not disappeared.

But let us begin by paying homage to what is irretrievably gone. In the striking words of Carolyn Merchant, "the world we have lost was organic." Until the scientific revolution, the central image of the earth was that of a nurturing mother, "a kindly, beneficent female who provided for the needs of mankind in an ordered, planned universe."[7] This nurturing mother gave birth to plants and animals, and ultimately to human beings. Streams, forests, and minerals also were her children. Minerals, in particular, were regarded as living organisms that grew inside the earth as an embryo develops in the uterus, gestating in the warm, dark, womblike matrices of subterranean space. It was widely assumed that minerals could grow and propagate. From antiquity to the seventeenth century, observers described mines of iron ore that supposedly refilled themselves, or silver mines where the ore grew in plantlike patterns on abandoned timbers. Often the metaphor of a golden tree was used to describe the origins and the distribution of minerals. Veins of mineral ore were seen as branches of an immense trunk that extended down into the deepest parts of the earth, where no mine could reach. There, it was said, metals were produced by Mother Earth and eventually rose, like sap in a tree, through cracks in the soil that corresponded to vessels in plants.[8]

According to this worldview, the earth was not a neutral resource to be exploited for human benefit; it was a sacred entity. To delve into the earth was akin to rape. Mining was therefore an enterprise of dubious morality, comparable to mutilation and violation. Up to the end of the Middle Ages in Europe, the sinking of a mine was a ritual operation. Religious ceremonies were held first, because the area that was going to be entered was sacred and inviolable. The domain that held the mysteries of mineral gestation did not, by right, belong to humankind.[9]

Merchant emphasizes the ethical consequences of this worldview. In her opinion, the sacred image of Mother Earth was so powerful that it discouraged (although it never prevented) mining. To be sure, there were many practical reasons for society to discourage mining, since mining activities had long been observed to entail unfortunate social and environmental effects. But Merchant argues that the dominant metaphor itself carried ethical restraints.

For mining to be carried out on a large scale, these restraints—and the metaphor of Mother Earth on which they were based—had to weaken. In Merchant's worlds, "Society needed these new images [of mastery and domination] as it continued the processes of commercialism and industrialization, which depended on activities directly altering the earth. . . ."[10] For example, she shows how Georgius Agricola (1494–1555), the foremost early writer on mining technology, took pains in *De Re Metallica* (published in 1556) to counter traditional ethical objections to mining. In response to the argument that minerals were hidden in the earth because they were not intended to be extracted, Agricola replied that just because fish swim below the water does not mean we should refrain from catching them.[11]

For all her admiration of the organic universe, Merchant's interpretation of the social universe leans toward the mechanistic. In the way she connects technological and ethical change, she implies a conventional Marxist interpretation of the relation between productive base and intellectual superstructure: as the technological base changes with the advent of capitalist commerce and large-scale industrialization in mining, the intellectual superstructure (metaphors, values, and norms) must change to accommodate the new economic activities. Lewis Mumford, who shares much of Merchant's historical outlook (and who also severely criticizes Agricola's instrumentalist values), would probably object to this implicit technological determinism. As Mumford stresses in *Technics and Civilization*, commercialism and industrialization should be seen as the *results* of cultural transformations—new interests, new standards of value, new concepts of time and space—as much as they are seen as having *caused* these changes. The development of mining technology did not simply generate a need for new cultural images; it came about in the first place because of preceding cultural developments. Moreover, the technological changes themselves provided new images. In particular, the new type of intellectual inquiry that emerged in the late Renaissance—then called natural philosophy and now called science—depended upon mining images to explain its principles and methods.

The key figure here, of course, is Francis Bacon (1561–1626). As Merchant and Mumford both stress, Bacon used aggressive metaphors of mining and smithing to describe the activist, intrusive principles of natural philosophy. Because the earth's womb harbors the deepest secrets of nature, Bacon contended, that was where the natural philosopher must seek truth: "There is therefore much ground for hoping that there are still laid up in the womb of nature many secrets of excellent use having no affinity or parallelism with anything that is now known. . . ."[12] Far from being awed by the prospect of entering nature's secret parts, researchers should dig "further and further into the mine of natural knowledge." Within earth's bosom "the truth of nature lies hid in certain deep mines and caves," which must therefore be penetrated to discover her secrets.[13]

As Bacon was well aware, he was proposing a stunning reversal of metaphors and values. For thinkers of antiquity, the essence of thought was contemplation, which alone permitted the philosopher to escape from the cave of deception and to emerge into the light of truth. For Bacon, on the other hand, truth was to be discovered not by contemplation but by action. The philosopher must descend from the daylight into the cave in order to mine its hidden secrets. Mining, a type of labor long regarded as worthy only of slaves, was now held up as a model for intellectual activity.[14]

For both Merchant and Mumford, then, Bacon represents the intrusive, aggressive, exploitative character of modern science. Both of them make a powerful critique of the scientific worldview by revealing its reliance on images of male domination and its denigration of female imagery and experience. (This consciousness of gender bias is not particularly strong in Mumford's *Technics and Civilization,* but it is found in his later writings.) One problem with this analysis, however, is that it tends to equate the rise of modern science with the triumph of utilitarian Baconianism. The project of modern science began with Bacon, but not only with him. Moreover, science has developed in directions that the pragmatic Bacon never imagined. Science unquestionably comprises an instrumentalist drive to master the nonhuman world and a view of nature as a mine to be excavated for utilitarian purposes. But science also includes an ecological tradition—a drive to understand the universe, with the assumption that nature has intrinsic values unrelated to human ones. To use the distinctly judgmental (and controversial) terms of the historian Daniel Worster, if there is an "imperialist" science there is also an "arcadian" one. Or, to call upon the more general analysis of Max Horkheimer and Theodor Adorno in *The Dialectic of Enlightenment,* the legacy of the Enlightenment is twofold: it includes the goal of subordinating a desacralized nature to human ends and also the goal of freeing the human mind by transcending the human-based view of nature and apprehending more fundamental sources of the natural order.[15]

That is one part of the problem with the Merchant-Mumford analysis: it tends to equate science with Baconianism—to consider science as a bloc rather than to recognize other, less dominant traditions. The other drawback of their analysis is that it tends to equate intention with result. Early students of the earth may indeed have intended to follow a rational Baconian program, throwing out old myths and organic analogies in favor of direct observation and experiment.[16] Certainly Bacon's followers in England kept reminding themselves and one another that, in order to avoid credulity, they must consult not books but the rocks themselves. (As one of Bacon's disciples, Edward Lhwyd, insisted, "with Natural History know there's no good to be done in't without repeated observations."[17]) In carrying out the Baconian program, however, investigators went deeper and deeper below the surface of the earth, and as they went deeper they discovered a world far older and far stranger than Bacon or his immediate disciples ever imagined. Excavation sought a rational past and uncovered a quasi-mythological one.

To understand how scientific investigation of the underworld veered off the Baconian track, we must look in more detail at the history of geological excavation. For at least a century after Bacon's death, his followers in Britain were extremely limited in their ability to see below the earth's surface. In the later 1700s the construction of canals and roads, and in the 1800s of railroads, would provide informative cuttings, but before then such projects were rare. The most readily available sources of information about the inner earth were rock outcroppings. They were not easy to interpret, however—especially in Britain, where many of them were covered with soil and vegetation.[18] Baconian naturalists such as Lhwyd had to be observant and indefatigable travelers in order to gather useful observations. They also distributed numerous questionnaires (Lhwyd sent out 4,000 for his survey of Wales) asking local observers, usually country gentlemen and parsons, for "subterraneal observations" on minerals, quarries, beds of stone, types of clay, and the like.[19] Later naturalists used even more detailed and sophisticated questionnaires to gather information on fossils and minerals, so that locations could be described both horizontally and vertically.[20]

According to Baconian philosophy, once such information was accumulated it was supposed to be classified. Catalogs of both minerals and fossils were published in England in the late 1600s and the early 1700s, as well as maps showing vertical columns of strata as deduced from mineshafts and wells, bird's-eye views of landforms, and sketches of striking rock formations. What was missing, though, was the concept of large-scale, regularly occurring strata or rock types. Although Agricola had done important work on stratification, and although Nicholaus Steno had formulated the basic principles of stratigraphy in the 1660s, their ideas lay undeveloped. To most naturalists, the earth's strata seemed jumbled and random. Minerals were still assumed to originate in local chemical combinations or special crystallizing conditions.[21] Though the order of the heavens had been triumphantly established, there still seemed to be no orderly system below the earth.[22]

Only late in the eighteenth century, in Central Europe, did researchers come to understand that the earth is composed of layers of strata laid down in a regular and predictable way. Unlike the English disciples of Bacon, the Central European investigators were in close contact with the mining industry. The mines in Bohemia, Saxony, and Hungary were larger, deeper, and technically more advanced than those in the British Isles; when mine operators in Britain needed help in locating minerals or in digging deeper pits, they hired consultants from Central Europe. In the eighteenth century, German scholars continued the tradition of Agricola and carried on mineralogical studies based on direct contact with the mining industry. Unable to convince the traditional universities to teach subjects related to their enterprise, leaders of the mining industry began in the 1760s to organize their own mining academies in Prague, Freiburg, Berlin, and other centers. The curricula included engineering, assaying, and mineralogy.[23]

The Freiburg Academy rose to preeminence, largely because Abraham Gottlob Werner (1750–1817) began teaching there in 1775. Werner's ancestors had been leaders in the mining and metallurgy industries of Silesia. From boyhood he had collected minerals of the region, which his father (an inspector in an ironworks) had identified for him. With inexhaustible energy and charm, Werner achieved fame as the outstanding geologist of his day. Werner is now best known as the loser in the great battle of the "Neptunists" and the "Vulcanists" that raged in the 1780s. Werner, a "Neptunist," theorized that the various features of the earth's crust had been laid down when a primitive ocean had subsided. From this debate, which continued into the nineteenth century, modern geology was born. The conflict of the two theories provided testable hypotheses which inspired research programs instead of haphazard data-gathering. Even though Werner's hypothesis of a universal ocean did not long endure, it called attention to the regularities in the earth's structure. Since a universal ocean would produce universal formations, Werner devised a fourfold classification system corresponding to four major periods of earth history (primitive or primary, transition, floetz or secondary, and newest floetz or alluvial). In his lectures Werner stressed that the layers of the earth were laid down in an orderly series, which he described with great accuracy in terms of a geological column with numerous subdivisions.[24]

By 1800 the concept of a universal stratigraphic column had been firmly established. The agreement on the concept was far more important than any arguments about its details.[25] The image of the column—vertical sections in the earth, corresponding to enormously long periods of time—became the central representation of deep time. In some respects the development of this image may be seen as a triumph for Baconianism, for it represented the achievement of classification after decades of observation. In other respects, though, the column opened the way to a recovery of the mythological. Researchers discovered that the most reliable way to read the geological record was to correlate it with the biological one. Fossils, which were by then recognized as organic remains, became the key to identifying strata—and this key unlocked a strange and wondrous world of past lifeforms.

The scientist who added "flesh of paleontology"[26] to the layers of rock was Baron Georges Cuvier (1769–1832). With his colleague

Alexandre Brongniart (a trained mining engineer and an experienced field geologist), Cuvier carried out extensive investigations in the Paris Basin in the early 1800s, especially in the many quarries in the area. Cuvier and Brongniart found that rock strata, even very thin ones, occurred in a predictable, fixed vertical order throughout much of the Paris Basin.[27] They were amazed to find that this constancy extended over 120 kilometers, and down to the thinnest layers. In these and many other excavations, Cuvier concluded that the most reliable way to identify strata was to identify key fossils. He became expert at comparative anatomy, especially at structural comparisons of living and fossil animals. Cuvier was the first to demonstrate that extinct animals could be reconstructed from fragmentary remains according to consistent anatomical principles; he thus reconstructed extinct hippopotamuses, rhinoceroses, cavebears, mastodons, crocodiles, and tigers.

Even stranger discoveries were to follow. In 1834 the British paleontologist Gideon Mantell used Cuvier's principles to reconstruct the iguanodon from bones found in chalk pits near his Sussex home. One of the many visitors who Mantell said "besieged" his house was the celebrated artist John Martin (1789–1854), upon whom Mantell prevailed to illustrate the first volume of his *Wonders of Geology* (1838). There Martin portrayed (again in Mantell's own words) "the appalling dragon-forms" of the dinosaurs in an engraving titled "the country of the iguanodon." In later books by other scientists Martin illustrated the ichthyosaurus and the plesiosaurus in images "founded on scientific data but, for all that, unmistakably, if distantly, related to the dragon Cadmus slew."[28]

The great problem of paleontology in the nineteenth century was to explain the relationship between extinct and extant forms of life. Cuvier recognized that the bones he had found were of animals that no longer existed, but he resisted any suggestion that one life-form could evolve into another. He believed that each species was fixed, and that the succession of different lifeforms was due to a succession of sudden and overwhelming geological catastrophes which had abruptly destroyed most species of plants and animals and allowed new ones to replace them. In brief, Cuvier, like Werner, is now best known for being wrong. Cuvier's error (according to standard accounts) was to lead the "catastrophist" school of geology, which eventually lost to the "uniformitarianism" espoused by Charles Lyell and his admirers, among them Charles Darwin.

Yet the debates between catastrophists and uniformitarians, as well as those between Neptunists and Plutonists, presumed agreement on this crucial point: mighty natural events in the past had laid down an orderly geological record that was also a chronological record. Whatever their other differences, geologists in the first third of the nineteenth century further agreed that a prime goal was to fill in that record by refining the stratigraphic column.

In 1815, William Smith (1769–1839), a surveyor and engineer who specialized in laying out canals, published a map of England that was far more detailed than any previous one. It measured nearly nine feet by six feet, and twenty colors, shaded from light to dark, were used to identify the succession of strata in each area. In subsequent years Smith published books to accompany the map, including one detailing the fossils that appeared in various strata.[29] About the time that Smith's map appeared, William Buckland (1784–1856) was appointed as the first reader in mineralogy at Oxford. A superb field geologist, Buckland published a comparative table of continental and English stratigraphy in 1821. During the 1820s the table continued to be filled in and improved. By 1830 standard wall charts of the stratigraphic column had appeared, and by 1840 such now-familiar terms as *paleozoic* had been introduced.[30] It was now understood that the scale of geological time had to be immensely long—long enough for seabeds to have become land and vice versa. In the first edition (1830–1833) of his famous book *Principles of Geology,* Charles Lyell took the dramatic step of presenting a stratigraphic table that was open at the bottom—a vivid image of his conviction there was no beginning point in the history of the earth.

In comparison with the stratigraphic measure of time, the much more familiar chronology of human history (Old Testament times, Greece, Rome, the Christian era, the Middle Ages, modern times) now seemed insignificantly brief. The great problem was to correlate the human timescale with the geological one. The obvious place to look for a splice was the Biblical story of the Flood, a great geological event that was also a great historical event in the Judeo-Christian tradition.

Cuvier proposed that Noah's flood was the last in a series of catastrophic *bouleversements* (of which Cuvier's disciples eventually enumerated 27). In these universal upheavals the land and the sea had exchanged places, so that no trace of antediluvian creatures, human or otherwise, would be found; any evidence would now lie at the bottom of the ocean. Cuvier visited the studio of John Martin when the artist was working on a second version of his famous painting *The Deluge,* and came away highly pleased. Martin's canvas showed day and night mingled, the earth opened up to engulf the unrighteous, caves become traps, the waters and the sky blended in a volcanolike tempest. Martin had painted the kind of global upheaval Cuvier envisioned.[31]

English geologists, foremost among them Buckland, also believed in a series of floods, but argued that they had been less catastrophic—more like a giant tidal wave than a universal *bouleversement.* The English geologists therefore assumed that evidence of antediluvian life might be found. It had long been known that caves sheltered the bones of species that no longer existed or no longer lived in northern climates. Would not caves have sheltered animals as they sought to escape rising flood waters? They were the logical place to look for evidence of antediluvian life, especially limestone caves where water dripping from the roof would cover any bones lying beneath with a protective coating of stalagmite.

Beginning in 1816, Buckland inaugurated modern cave research by exploring some caves in the German region of Franconia. The new method of excavation stressed finesse, not force. Each layer of soil, each stalagmite, each scrap of evidence had to be carefully recorded if convincing evidence of past life forms was to be assembled. Buckland's techniques received their severest test when, in the fall of 1821, he was notified of the opening of a cave in Kirkdale, Yorkshire. Because the Kirkdale cave appeared to be unusually rich in a wide variety of fossil carnivores, it promised to provide the evidence of the authenticity of the Flood. Buckland excitedly wrote a friend that only "a common calamity of a simultaneous destruction could have brought together in so small a compass so heterogeneous an assemblage of animals."[32] In December 1821 Buckland visited the site, examined it thoroughly, and collected specimens. His work was masterly. From a seemingly haphazard heap of teeth and bones, using the principles of Cuvier's comparative anatomy, Buckland identified both carnivorous and herbivorous animals, both extinct and existing ones, both tropical animals and northern ones. Furthermore, he combined his paleontological work with observations of contemporary animals (hyenas in an Exeter menagerie) to produce a detailed and careful analysis of how these animals crush and crack the bones of their prey. To his credit as a scientist, Buckland concluded that the evidence did not support his initial theory of a "common calamity." Instead, he reported, the cave had been a den of antediluvian hyenas, whose eating habits explained why the bones were piled up as they were. Buckland's report was published in *Philosophical Transactions* and in other scientific and popular journals in Britain and on the continent, and for his work he was honored with the Copley Medal of the Royal Society.[33]

Buckland by no means gave up his conviction that caves harbored evidence of the Flood, however. In 1823 he published the much-read, much-reviewed, much-admired book *Reliquiae Diluvianae* [Relics of the Flood], the contents of which are best summarized by the subtitle: *Observations on the Organic Remains Contained in Caves, Fissures, and Diluvial Gravel, and on Other Geological Phenomena Attesting the Action of a Universal Deluge*. Buckland cited detailed evidence from caves, from layers of gravel and loam, and from the shapes of hills and valleys to conclude that all these geological formations resulted from the dispersal of surging floodwaters.[34]

But one crucial piece of evidence was missing: human fossils. If antediluvian creatures had taken refuge in caves to escape the rising waters, why were the sinners who had incurred God's wrath not among them? The absence of human fossils had not been a problem for Cuvier. According to his catastrophic theory, any such remains would now lie at the bottom of the deepest ocean, deposited there when earth and sea changed places. Cuvier and his disciples flatly declared "L'homme fossile n'existe pas."[35] According to Buckland's theory, though, caves should hold the remains of the human beings who were, except for Noah and his family, swept away in the Flood. When he began his cave research, Buckland was optimistic that such evidence would be discovered. In a note dated 1819 he wrote: "The examination of diluvian gravel of Central and Southern Asia would probably lead to the discovery of human bones which is almost the only fact now wanting in the series of phenomena which have resulted from the Mosaic Deluge."[36]

When evidence of antediluvian human life was unearthed, however, Buckland was not convinced by it. For example, when a skeleton and related stone artifacts were found in Goat Hole Cave (a sea cave near Paviland, South Wales), Buckland stated in *Reliquiae Diluvianae* that the skeleton was relatively recent—possibly dating from Roman times, when caves had often been used as sepulchers. He was also skeptical when in 1824 a Catholic priest, Father Mac-Enery, found flint and bone tools in Kent's Cavern, near Torquay on the south coast of Devonshire. These man-made artifacts were found alongside teeth and bones from the same extinct animals Buckland had found near Paviland; in Kent's Cavern, furthermore, both tools and bones were sealed under a floor of limestone deposited by dripping water. Here, it seemed, was proof that man had lived in England at the time of the mammoth and the cavebear. Buckland insisted, however, that "Ancient Britons" inhabiting the upper levels of the cave had scooped out "ovens" in the stalagmite and that their tools had fallen down through those holes. In the face of Buckland's opposition, Father MacEnery decided not to publish his records.[37]

Cross-section and ground plan of the Welsh sea cave called Goat Hole, showing the location of the human skeleton found there. Drawn by T. Webster from a sketch made by William Buckland. From Buckland's *Reliquiae Diluvianae*, second edition (London, 1824). Courtesy of the Bernhard Kummel Library of the Geological Sciences, Harvard University.

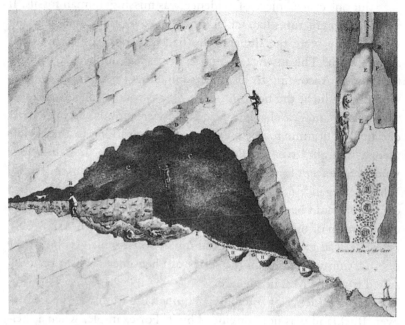

Later in the 1820s, in the vicinity of Liège, Philippe-Charles Schmerling methodically explored forty caves along the banks of the Meuse, finding cemented together in the floors of the caves the bones of extinct animals, implements made of flint, ivory, and bone, and a human skull. Buckland visited Schmerling but was still unconvinced. He returned to England, where he continued to study Celtic and Roman antiquities in an attempt to gather evidence for his sepulchral theory.[38]

As a prudent scientist, Buckland had good reason to question these finds. Earlier researchers had embarrassed themselves by carelessly accepting poor evidence. (Cuvier had shown that one specimen of "*Homo diluvii testis*" was in fact the remains of a large salamander.[39]) But Buckland's skepticism was so persistent and his reasoning (at least in the case of Kent's Cavern) so convoluted, that obviously he was of two minds with regard to the prospect of finding human fossils. Part of him resisted the idea of man's great antiquity. It had been a blow to humanity's view of its own central role in creation to read the geological record back into the dim and distant past, down a long and maybe even bottomless stratigraphic column. It was also unsettling to discover a multitude of bizarre animals that seemed to have existed long before human life. Most unsettling of all was to imagine that humanity too was immeasurably old, and that there might be extinct forms of human life, recoverable only as fossils.

Buckland held to the uniqueness of humankind as the "non-fossilized apex of progressive change."[40] In 1836, in his contribution to the widely read Bridgewater Treatises, Buckland gave up the idea that the last great geological flood had been the same as the scriptural one. The Mosaic event, he commented in a footnote, must have been relatively minor. By giving up the idea that the Flood had been a unique event, Buckland was able to keep the splice between geological and human history at a fairly recent point; thus he was able to reject the idea of human fossils.

Buckland's skepticism about human fossils was shared by many. Charles Lyell, the great defender of uniformitarianism, also visited Schmerling in Liège in 1833 and also came away unconvinced.[41] Similar disbelief met Jacques Boucher de Perthes, a French customs officer, when in the 1830s he began to collect and catalog the animal bones and stone tools (among them an axe buried in an antler haft) that had been hauled up by a dredger working in the river near Abbeville. When he submitted his book *Antiquités celtiques et antédiluviennes* to the Institut de France in 1846, Boucher de Perthes was ridiculed. The geologist Élie de Beaumont scoffed at the unknown provincial and said that the flint implements were from Roman times; man, he asserted, could not have lived in France at the time of the elephants.[42]

In 1858 an intact bone cave was discovered when its roof fell in during quarrying operations at Brixham, Devonshire (a few miles west of Torquay). This time geologists were able to perform a scrupulously careful excavation. A distinguished committee that included Charles Lyell, the paleontologist Hugh Falconer, and the geologist Joseph Prestwich was set up to sponsor the work.[43] In this case there could be no serious argument about the results: unmistakable flint knives lay in the floor of the cave below a bed of stalagmite in which the bones of mammoths, cavebears, and other extinct animals were encased.

In the same year of the discovery of the Brixham Cave, Hugh Falconer visited Jacques Boucher de Perthes in France. He was so impressed by the evidence of the Abbeville gravels that upon his return to England he persuaded Joseph Prestwich to go see for himself. Prestwich did so, making the visit with John Evans in May 1859. In June, after their return to England, Evans presented their findings to the Society of Antiquaries, and Prestwich read a paper to the Royal Society summarizing the evidence from Amiens, Abbeville, Brixham Cave, and Kent's Cavern (the latter had been excavated again in 1846, without any trace of Buckland's "ovens" being found). In the audience were Charles Lyell, Thomas Huxley, and Michael Faraday. Lyell shed the doubts that had led him to question Schmerling's discoveries thirty years before, and used his prestige to convince scientists and the public that the evidence was genuine.

The year 1859 also brought the publication of Darwin's *Origin of Species*. Though silent on the specifics of human evolution, Darwin's book implied descent from the apes in remote times. Just two years before the *Origin* appeared, workmen in a quarry near the Prussian town of Neanderthal (about 70 miles northeast of the Liège caverns excavated by Schmerling) had found a human skeleton. The skeleton had been scattered by the workmen, its age was a mystery, and no animal remains were found with it. Most baffling of all was the shape of the skull—a low narrow forehead, large projected ridges above the eyes, and thick cranial bones.

When Lyell presented a cast of the Neanderthal skull to Huxley, the latter "remarked at once that it was the most apelike skull he had ever beheld."[44] Huxley's widely read 1863 essay *Man's Place in Nature* included a long account of the Neanderthal skull aimed at refuting those who claimed it was that of a congenital idiot or, possibly, a Cossack or "a powerfully organized Celt" (since it "somewhat resembl[ed] the skull of a modern Irishman with low mental organization"[45]). This was, Huxley said, the skull of an apelike human. He began the essay with these words:

Ancient traditions, when tested by the severe processes of modern investigation, commonly enough fade away into mere dreams; but it is singular how often the dream turns out to have been a half-waking one, presaging a reality. Ovid foreshadowed the discoveries of the geologist: the Atlantis was an imagination, but Columbus found a western world: and though the quaint forms of Centaurs and Satyrs have an existence only in the realms of Art, creatures approaching man more nearly than they in essential structure, and yet as thoroughly brutal as the goat's or the horse's half of the mythical compound, are now not only known, but notorious.[46]

They became even more notorious as other humanlike fossils were found. In 1868, in the course of constructing a railway line, a rock shelter in a cliff bordering a river near the French village of Les Eyzies was blasted open. In this cave were found four skeletons similar to the one Buckland had examined from the Paviland cave: tall, large-brained, long-headed. These Cro-Magnons (as they became known) had apparently been formally buried with pierced shells and other ornaments. As additional bones were found, it became evident that the Cro-Magnon people had succeeded the Neanderthals. Even more startling evidence of the cultural achievements of the Cro-Magnons came in 1864, when a piece of ivory sharply engraved with a lively drawing of a hairy mammoth was found in La Madeleine, a cave in the Dordogne Valley of France. In the 1890s, in the same valley, cave paintings of superb quality were found. Other cave art had been found in Spain in 1879, but then few had considered it authentic. In 1902, however, a group of leading French archaeologists visited the Dordogne caves and solemnly declared that the paintings there were the work of ancient humans.[47]

In 1865, with his widely read book *Prehistoric Times,* Sir John Lubbock (later Lord Avebury) popularized the terms *prehistory* and *prehistoric.*[48] Lubbock proclaimed on the first page of *Prehistoric Times* that "of late years a new branch of knowledge has arisen, a new science has . . . been born among us, which deals with times and events far more ancient than any which have yet fallen within the province of the archaeologist. . . . Archaeology forms the link between geology and history."[49] The Neanderthal skull played a pivotal role in the development of archaeological science. Like Galileo's telescope in the Copernican revolution, it provided direct, powerful, sensory demonstration of a new science that denied man's central place in creation. The image of that heavy-browed, heavy-jawed skull spoke more eloquently than any words about man's remote origins among the primates. Cave exploration, undertaken to confirm Biblical history, instead uncovered a quasi-mythological world of dragonlike beasts and satyrlike humans.

In describing the history of archaeology, Jacquetta Hawkes makes a distinction between the "scientific" and the "humanistic" archaeology of the nineteenth century. The former unearthed humanity's prehistoric origins; the latter uncovered its known civilizations. The recovery of ancient civilizations through humanistic archaeology was far more gratifying to Westerners' self-esteem than the unearthing of apelike skulls. In an age when higher education meant studying the classics, many Europeans were profoundly excited to see archaeologists literally digging up the cultural history of the West.

The term *humanistic,* however, tends to mask the militaristic and imperialistic character of this type of archaeological excavation. Greed, nationalism, and cultural pride, more than science, motivated the excavation of the buried worlds of the Middle East. As much as mining operations, much of the so-called humanistic archaeology of the nineteenth century exemplifies the plunder of the earth.

The origins of modern archaeology are often traced to Napoleon Bonaparte's invasion of Egypt in 1798. When he embarked from Toulon in that year, Napoleon took with him a scientific staff, which eventually became the French Institute of Cairo. The next year, while French soldiers were digging the foundations of a fort near Alexandria, they found the piece of basalt that would become known as the Rosetta Stone. These two events mark the beginning of modern

Egyptology. The French Institute remained in Cairo when the French evacuated in 1801 (leaving most of their collection of portable antiquities, including the Rosetta Stone, to fall into the hands of the British). From 1809 to 1813 the institute published an important series of volumes, the *Description de l'Égypte,* and in 1822 J.F. Champollion used the Rosetta Stone to decipher Egyptian hieroglyphics.

By then, the excavation of ancient Egypt was beginning. The body of the Great Sphinx (which had been covered with sand, so that only the head was exposed) was uncovered in 1818. In the 1820s and the 1830s, various expeditions were at work in the Valley of the Kings, discovering new pyramids (bringing the total to 67 by the 1840s) and clearing away the sand from the Great Temple at Abu Simbel. These early digs consisted mainly of looting the larger, less fragile objects, such as stone sculptures. Many smaller and more delicate items were destroyed in the process. In 1837, Colonel Howard Vyse used gunpowder to blast entrances into some of the pyramids.[50] Giovanni Belzoni, working in Egypt under the direction of the British Consul-General, described how upon opening one tomb he ended up in a pile of broken mummies: "I could not avoid being covered with legs, arms and heads rolling from above."[51] The general pillage continued until 1850, when Auguste Mariette, sent by the Louvre to head the Egyptian Antiquities Service, helped protect Egypt against further looting by founding the Cairo Museum.

Much the same happened in Mesopotamia. In the 1840s the Frenchman Paul-Émile Botta and somewhat later the Englishman Austen Henry Layard dug into various mounds dotting the region in an effort to find the ancient Assyrian capital of Nineveh. Both were convinced they had exhumed Nineveh,[52] and they raced to unearth its treasures. At one point Layard's assistant, digging secretly after dark at a corner of the mound where the French were already at work, discovered the palace of Ashurbanipal and the "lion hunt" sculptures now in the British Museum. Layard's finds were spectacular. He went on to excavate large parts of Ashurbanipal's clay-tablet library, the black obelisk of Shalmaneser III, and a number of huge man-headed, winged bulls and lions.

Because of his impatience, his eagerness to collect museum exhibits, and his inadequate funding, Layard used methods that were recognizably destructive even in his own day. He tunneled through palaces and temples, filled up trenches with rubbish from subsequent cuttings, and generally left "a terrible hotchpotch for his successors."[53] But the audience back home was enthralled. From 1847 on, a *Morning Post* correspondent was on the scene in Assyria, publishing regular dispatches. In 1849, upon returning from his first expedition, Layard wrote a book called *Nineveh and Its Remains,* which sold briskly. In France, some of the earliest published books of photographs showed archaeological sites and monuments in the Near East. Motifs from ancient Egypt and Assyria become decorative and architectural clichés, adorning everything from railroad stations and tunnel entrances to middle-class mantelpieces.[54]

In the 1870s came a discovery that did even more to revive the semi-mythological past: Heinrich Schliemann's discovery of the site of Homer's Troy. Schliemann was by no means a pioneer in geo-archaeology, and his techniques have been severely criticized; nonetheless, his work represents a significant advance over that of Layard and Belzoni.

In Mesopotamia, where brick is the universal building material, archaeologists often face stratification as complex and confusing as that confronted by geologists. When buildings are abandoned the roofs collapse, the walls crumble, and the room fills up; when new occupants decide to rebuild, they level these remains and start again, somewhat higher. In addition, heavy winter rainstorms help to weld standing walls and fallen rubbish into "such a homogeneous mass that only a practised eye can distinguish one from the other."[55] Amid such difficulties, Schliemann carefully kept all his finds (even the seemingly unimportant ones), supervised his workers carefully, and tried to record every detail.[56] He initially identified seven levels of habitation, and when he was later joined by a more skilled excavator they distinguished nine.[57]

In the late 1870s and the 1880s, humanistic archaeological excavation began to approach the precision and thoroughness of the scientific variety. In 1875 the German Archaeological Institute founded a branch at Athens and took charge of excavations at Olympia; many regard its work as marking the beginning of the use of modern scientific techniques in archaeology.[58] The Englishman William Flinders Petrie, who arrived in Egypt in 1880 (he was to stay there until 1935), introduced far more rigorous standards there than had ever been used before. He recorded carefully, excavated methodically, and published promptly and completely. Petrie also devised

The excavation of the Great Tower of Ilium, as supervised by Schliemann. The top of the tower, which is 20 feet high, lies 26 feet below the surface of the hill. From Schliemann's *Troy and Its Remains* (London, 1875). Courtesy of Lamont Library, Harvard University.

the system of sequence dating, by which he fixed the relative dates of finds on the basis of stratification and on the basis of successive forms of pottery—the archaeological equivalent of dating strata by fossils.[59] A third pioneer in scientific archaeology was General Augustus Henry Pitt-Rivers, who worked primarily in Dorset, where he had inherited an estate in 1880. With meticulous care he organized and carried out the excavations of entire villages, counting the houses and estimating the number of inhabitants and the amount of food they had available.[60] The development of these techniques opened up the golden age of archaeology—the period that saw Sir Arthur Evans's work at Knossos, the discovery of Minoan Crete, Leonard Woolley's discovery of the royal graves at Ur, and Howard Carter's discovery of the tomb of Tutankhamen.

The fact that archaeology had become much more scientific did not mean that it had become a coldly rational enterprise. Modern archaeologists continued to experience many of the emotions that Carolyn Merchant and Mircea Eliade attributed to preliterate miners: the sense of intruding into sacred mysteries, the feeling of awe and even reverence. Even the impulsive and impatient Layard, recalling how he watched the first winged bull emerge from the soil of Assyria, said: "It required no stretch of the imagination to conjure up the most strange fancies. This gigantic head, blanched with age, thus rising from the bowels of the earth, might well have belonged to one of those fearful beings which are described in the traditions of the country as appearing to mortals, slowly ascending from the regions below."[61] And here is Howard Carter describing the day (November 26, 1922) when, for the first time, enough debris was removed from the doorway of Tutankhamen's tomb so that, with a candle and then a flashlight, he could peer inside: ". . . as my eyes grew accustomed to the light, details of the room within emerged slowly from the mist, strange animals, statues, and gold—everywhere the glint of gold. For the moment—an eternity it must have seemed to the others standing by—I was struck dumb with amazement, and when Lord Carnarvon, unable to stand the suspense any longer, inquired anxiously, 'Can you see anything?' it was all I could do to get out the words, 'Yes, wonderful things'. . . ."[62]

The idea that excavation is a solemn task, on the fringes of permissibility, had not disappeared. Carter commented: "I suppose most excavators would confess to a feeling of awe—embarrassment almost—when they break into a chamber closed and sealed by pious hands so many centuries ago. . . . Time is annihilated . . . and you feel an intruder."[63] This sense of awe also affected the public at large, especially when many of those who had participated in the Carter excavation died suddenly and unexpectedly. (The world murmured about the curse of King Tut's tomb.) In its seeming infinity, without discernible beginning or end, deep time retained an aura of mystery and sacredness. Geology and archaeology have altered but not ended the human capacity for shaping meaning from subterranean space.

And as imagination reached further into the past, it also leapt further into the future. The sight of the ruins of past civilizations inevitably suggested what the present would be in the future, when what was now on the surface would become part of the buried past. As early as 1817, Shelley had written that around the mighty Ozymandias, king of kings, "the lone and level sands stretch far away." The rest of the nineteenth century was haunted by that prospect of future burial. Mary Shelley imagined the end of the world in *The Last Man* (1826), and subsequent British artists portrayed this haunting figure wandering among half-buried misty temples or (in a watercolor by John Martin) standing high on a cliff overlooking a desolate abyss.[64] Other painters showed the rotunda of the Bank of England lying in ruins, half-covered with dirt and weeds. In France, Victor Hugo's poetic cycle *À l'arc de triomphe* envisioned only three monuments—Sainte-Chapelle, the Vendôme column, and the Arc de Triomphe—remaining as traces of a vanished Paris. A similar vision inspired the masterwork of Maxime du Camp, who had traveled widely among the buried cities of the Near East. One afternoon in 1862, when visiting an optician's shop near the Pont Neuf—an errand that led him to meditate sadly on his own aging—du Camp resolved to write the kind of book about Paris that he wished historians of antiquity had written about their cities. The result was the fascinating multivolume study *Paris, ses organes, ses fonctions et sa vie dans la second moitié du XIXe siècle.*[65]

In literature about imaginary underground worlds, the association between depth and time was pushed even further, to greater depths and remoter epochs. In Jules Verne's *Journey to the Center of the Earth* the explorers find giant sea creatures and huge, vaguely prehistoric human figures in the innards of the planet. In Verne's *Twenty Thousand Leagues Under the Sea,* the submarine *Nautilus* glides

Howard Carter opening the doors of the second shrine within the sepulcher of Tutankhamen. Four shrines were nested in the burial chamber. Photograph by Harry Burton of the Metropolitan Museum of Art. Courtesy of Griffith Institute, Ashmolean Museum, Oxford, England.

by the site of ancient Atlantis. The three adventurers of Arthur Conan Doyle's *The Maracot Deep* (1927) contact the descendants of the Atlanteans when they venture into an ocean trench in a "little pressure-proof look-out station."[66] In Henry Rider Haggard's *When the World Shook* (1919), the adventurers unwittingly revive two survivors of an unnamed civilization that had sunk beneath the ground eons ago in a huge Cuverian cataclysm. The tales of Doyle and Haggard are markedly similar, each with a trio of male adventurers, a beautiful young woman, a demonic old man, and a civilization that sank below the surface in punishment for its evils. They tell of quests for the ultimate truth about time, and they suggest that the deepest time is in fact a circle. Their common theme is reincarnation: each of the male heroes discovers that he was part of the ancient past he has uncovered, that he loved the heroine in his past life, and that he is destined to love her again.

The rediscovery of cyclic time in the books of Doyle and Haggard is part of their more general spiritualist critique of modern science. Like many who were intrigued by occultism in the late 1800s and the early 1900s, they did not claim that modern science was false; rather, they claimed that it was inadequate, that its materialist premises could not account for many natural but mysterious phenomena. In *The Maracot Deep* Doyle makes a point of having his hero comment, when he first encounters the odd-looking descendants of the Atlanteans, "Already I began to surmise that no infraction of natural law was involved in the life of these strange people. . . ."[67] Haggard and Doyle were protesting not against natural laws, but against the reduction of those laws to ones governing matter in motion. They wrote in a scientific age still imbued with the Newtonian image of hard bits of matter moving through empty space and with the view of time as a separate, immaterial entity. Because time was invisible and noncorporeal, it seemed to them inherently a more "spiritual" entity than "materialist" matter. They preferred to criticize the mechanistic bias of science by exploring time rather than matter.

In fiction written from this point of view, those who venture into subterranean or submarine regions are both time travelers and spiritual pilgrims. They plunge below surface material reality to the truth that lies hidden below. The lower world is, paradoxically, identified with higher truth. This identification is the thread that connects Poe ("Reynolds! Oh Reynolds!"), Bulwer-Lytton (whose

The Coming Race was only one of several influential occultist novels he wrote),[68] Haggard (whose novels *King Solomon's Mines* and *She* also have subterranean and immortalist themes), and Doyle. Doyle's hero Dr. Maracot, after being visited by a spirit who gives him the power to subdue the demon Wanda, exclaims: "That it should have happened to me! . . . To me, a materialist, a man so immersed in matter that the invisible did not exist in my philosophy. The theories of a whole lifetime have crumbled about my ears. . . . It is only when you touch the higher that you realize how low we may be among the possibilities of creation."[69] In the late-nineteenth-century revival of occultism, the scientific worldview seems have swallowed its tail, joining again to its mythological origins.

The spiritualist convictions that surface reality can be misleading and that truth is found only by descending to uncharted depths are by no means limited to a few eccentric writers. Indeed, these convictions are central assumptions of modern intellectual inquiry. In history, economics, psychology, and linguistics—in the widest possible range of disciplines—the process of excavation has become the dominant metaphor for truth-seeking. The modern quest for knowledge is framed in the same spatial construct as the ancient quest for buried secrets.

In the development of modern intellectual inquiry, metaphorical and literal excavation have proceeded interactively. When Francis Bacon said that "truth is to be sought in the deepest mines of nature," he was speaking metaphorically. The image of digging deep into nature, of mining her secrets, is one of several key images Bacon used (another is that of torture) to express the necessity of interrogating nature in an active way. But excavation was not only a metaphor for Bacon's followers, who undertook actual excavation to discover the structure of the earth. As the historical sciences developed, investigators literally unearthed lost geological ages, lost civilizations, lost ages of man.

As a result of these findings, the study of history burst out of the tidy framework provided by the Bible and the classics. A striking example of the connection between literal and metaphorical excavation in historical studies is found in *Les Misérables,* where Victor Hugo constantly alternates between dramatic narrative (the story of Jean Valjean) and historical narrative (the story of the French people).

In the dramatic narrative, crucial events are literally set in the sewers. In the historical narrative, Hugo explains why he must descend there metaphorically. He devotes a chapter to rescuing from oblivion *argot,* the language of misery, explaining:

The historian of morals and ideas has a mission no less austere than that of the historian of events. The latter has the surface of civilization, the struggles of the crowns, the births of princes, the marriages of kings, the battles, the assemblies, the great public men, the revolutions in the sunlight, all the exterior; the other historian has the interior, the foundation, the people who work, who suffer, and who wait, overburdened woman, agonizing childhood, the secret wars of man against man, the obscure ferocities, the prejudices, the established iniquities, the subterranean reactions of the law, the secret evolutions of souls, the vague shudderings of the multitudes, the starvation, the barefoot, the bare-armed, the disinherited, the orphans, the unfortunate and the infamous, all the specters that wander in darkness. . . . Is the underworld of civilization, because it is deeper and gloomier, less important than the upper? Do we really know the mountain when we do not know the cavern?[70]

For the past century and a half, historians have responded to Hugo's challenge, digging beneath surface manifestations to unearth submerged groups (homosexuals, criminals, women), submerged evidence (dreams, sexual customs, mental constructs), and submerged forces (economic, technological, ecological).

Hugo's contemporary Karl Marx insisted even more radically upon the primacy of subsurface history. In a pivotal chapter of the first volume of *Capital* ("The Labor Process and the Valorization Process"), Marx turns from topics of commodities, exchange, and money and invites his reader into the inner sanctum of capitalism: "Let us therefore . . . leave this noisy sphere, where everything takes place on the surface and in full view of everyone, and [enter] into the hidden abode of production, on whose threshold there hangs the notice 'No admittance except on business.' Here we shall see, not only how capital produces, but how capital is itself produced."[71] The precise character of the historical substructure (Can it be defined as "technology"? Just what does Marx mean by "forces of production"?), and its precise relation to the historical superstructure (Does the lower level "determine" the upper?) have been debated endlessly.[72] Beyond debate, though, is Marx's passion to take his reader underneath the level of monetary and commodity exchange and to illuminate the previously hidden processes of production as the crucial forces shaping modern society.

For Sigmund Freud, too, the subterranean metaphor is central. The superego, the ego, and the id are the strata of the mind. To describe the relationship between the conscious and the outside world, Freud compares consciousness to the surface of the skin, which can develop a "crust"—a nearly inorganic shield protecting the personality against disruptive stimuli. Below this surface crust, the underlying layers go on living and responding to the stimuli that manage to penetrate to their depths.[73] As with Marx, with Freud the composition of the buried layer (the unconscious) and its relation to surface phenomena (the conscious) are open to debate, but once again the primacy of the buried layer is indisputable. In both Marxism and psychoanalysis, the moment of truth comes when the buried is uncovered: when the analyst taps the subconscious, when the proletariat seizes the forces of production. Both Marx and Freud depend so much upon subterranean imagery that it is now virtually impossible to read a text about the underworld without filtering it through a Marxist or Freudian interpretation—without reading the buried world as the subconscious, or the working class, or both.

The premise of structuralism, too, is that beneath cultural forms lie invariant structures, largely preliterate, from which derive the surface phenomena of cultural life. The linguistic philosopher Ferdinand de Saussure, one of the originators of structuralism, assumed that phonemes are the substratum of language and that the task of the cultural critic is to dig down to that level. In Saussure's case, the metaphor of stratification evidently derives from the late-nineteenth-century thinker Hippolyte Taine, who is now little remembered but who was a major intellectual influence in his day. In his then-famous triad of *race, milieu,* and *moment,* Taine assumed that *la race* (the great "interior" force) and *le milieu* (the "exterior" one) interact to create *le moment.* Saussure appears to have borrowed his preference for scientific analogies in general, and for geological layers in particular, from Taine.[74]

Saussure's distinction between "external" and "internal" linguistics has been echoed in Noam Chomsky's distinction between "surface structures" and "deep structures," the latter being identified by Chomsky as generative grammar.[75] Similarly, Claude Levi-

Strauss has applied vertical imagery to historical and social phenomena by assuming that deep unconscious structures underlie institutions and customs. Literary structuralists, such as Yury Lotman, have sought to isolate "deep" structures of a story, which are not apparent on the surface.[76]

The poststructuralists Jacques Derrida, Roland Barthes, and Paul de Man start by questioning the possibility of finding any substratum, any underlying origin or end of language. Without one, however, they have difficulty explaining "how deconstructionism itself can operate as an intellectual procedure without falling victim to the same processes of false consciousness it purportedly criticizes in others."[77] Derrida affirms that the project of deconstructionism is simply to define the limitations of concepts, to outline their historical closure. However, he also holds out the possibility that by digging far enough into language we might find "the crevice through which the yet unnameable glimmer beyond the closure can be glimpsed."[78] Deconstructionists do not reject the project of digging below the surface; rather, they deny that there is any firm substructure to be reached. Barthes speaks of a language "without bottom," of an open-ended process in which the criticism does the structuring, endlessly undermining itself[79]—a bottomless abyss that is the linguistic equivalent of Lyell's open-ended stratigraphic column.

Since the nineteenth century, then, excavation has served as a dominant metaphor for truth-seeking. The assumptions that truth is found by digging, and that the deeper we go the closer we come to absolute truth, have become part of the intellectual air we breathe. In this respect scientific inquiry retains an aura of the mythological, since the heroic quest for scientific truth has the pattern of a descent into the underworld. If we shift from the metaphorical to the literal level, we still find mythological overtones to the scientific enterprise. Two centuries of scientific excavation after Bacon's death revealed a past of gigantic reptiles, buried cities, fabulous treasures, and apelike humans. As Huxley noted, this is the stuff that dreams are made of. Both in its process and in its results, then, both in the enterprise of digging into subterranean spaces and in what it has found there, modern science acts in an enchanted world.

There is no necessary, inescapable relation between the progress of scientific understanding and the metaphorical death of nature, if we define the latter as the destruction of our imaginative response to nonhuman life. The discovery of deep time has forced the human imagination to grow, not to diminish. The advancement of scientific knowledge by no means destroys the sense of awe that comes from intruding into realms of lost time. And although the scientific world-view may demand that we distinguish metaphorical from factual responses to the earth's depths, it does not demand that we give up our need to create meaning through metaphor.[80]

The far more serious threat to nature's life comes from technological progress, at least as progress has been generally understood in modern times. Here the danger is not metaphorical but objective: the possibility that forests, ponds, oceans, soils, species, even entire ecosystems will be destroyed. If ecological destruction proceeds far enough, nature will die metaphorically too. Nature cannot be alive as a matter of value unless it is alive as a matter of fact. Our primary concern, then, should be to prevent the objective death of nature. For that we need not less science but more—at least, more science of the "arcadian" variety. We also need to be more critical of the assumptions about technological progress that are embodied in the feats of excavation described in the next chapter.

Excavations II: Creating the Substructure of Modern Life

3

According to Lewis Mumford, mining was a prime causal agent in the emergence of modern industrial capitalism. Beginning in the sixteenth century "the methods and ideals of mining became the chief pattern for industrial effort throughout the Western World." Not only the practical techniques but also the mindset (or animus, to use Mumford's term) of modern industry emerged from the mine to ripple through society. Like warfare, the other crucial agent of modern industrialization, mining depended on destruction and brute force. The ore had to be pounded, picked, broken, and melted. Entire forests were cut down to furnish wood for timbers, machines, and smelting. Mining furthermore encouraged the typically capitalist notion of value, which is abstract and quantitative and which depends on rarity rather than life-value. It therefore established a pattern of environmental and human degradation—in Mumford's words, "the pattern for capitalist exploitation."[1]

We have already seen how in *Technics* Mumford treats the mine as a metaphor of the technological environment and of the scientific worldview. Here he deals with its significance in nonmetaphorical, objective terms. The role of subterranean technologies in industrialization is commonly neglected because the artifacts themselves are hidden from view. Walt Whitman proclaimed

The shapes arise! . . .
Shapes of factories, arsenals, foundries, markets,
Shapes of the two-threaded tracks of railroads. . . .

But the shapes were also sinking, creating a manufactured under-ground world of grids and networks. The cosmos of modern tech-nology, as much as that of ancient mythology, has a vertical structure. As it reached upward in the shapes of skyscrapers, railway bridges, oil rigs, and missiles, it also sank into the earth in building foundations, railway tunnels, oil wells, and missile silos. The descent below the earth's surface was in part a quest for scientific truth; it was also a quest for technological power. The triumphs of modern industrial and urban life arise from connections buried below the surface of the earth. These structures rest on hidden infrastructures.[2]

The subterranean basis of modern industry began to be built between the late 1700s and the late 1800s in the form of a transpor-tation network of canals and railroads. This was the distributive system that supported the new productive system. The second step came in the last half of nineteenth century with the creation of networks to support "the metropolis of the industrial era"[3]: sewers, water mains, steam pipes, subways, telephone lines, electrical cables. As Mumford pointed out in early drafts for *Technics,* in the nineteenth century city planning began to involve not only the disposition of the surface but also an "underground system of functions [that] form as it were the physiological apparatus of the new city. . . . the mod-ern city plan involves a co-ordination of the super-surface city with the sub-surface city."[4] As some more recent historians of technology say, the modern metropolis is a "networked city."[5]

The installation of the infrastructure meant that, for the first time in history, excavation became a part of everyday life. Mining came out of the hinterlands into the heart of the city. Under the very feet of pedestrians throughout the United States, Britain, and Eu-rope, the earth was being dug up—not just once, but over and over—as various systems were installed. Beginning in the mid-1800s, the sight of men digging with spades and shovels, tunneling with picks and drills, shoring up cuts with timber walls, and building an un-derground maze of stone passageways and intersections and arches became more and more familiar. Between 1852 and 1865, when the English artist Ford Madox Brown painted an enormous canvas on the theme of "Work," he chose as his subject four muscular navvies, with splendid bodies and picturesque clothes, digging a deep ditch in the middle of a Hampstead street. For Brown, and for his Victorian audience, digging had become the generic image of labor. In the

The painting "Work" (1852–1865) by Ford Madox Brown. Courtesy of Manchester City Art Galleries, Manchester, England.

painting, elegantly dressed strollers pass by, a beggar holds a basket of wildflowers, and the social critics Thomas Carlyle and Frederick Denison Maurice look on—but the navvies are at the center of the composition. "They provide the foundation that sustains all other ranks."[6]

As the new material foundations of industrial and urban life were being laid, so were new social foundations. Technological networks are also human ones. The undermining of familiar surface patterns with unfamiliar subterranean ones was vivid evidence that old ways of life were being undermined. "The physical experience of technology mediated consciousness of the emerging social order; it gave a form to a revolutionary rupture with past forms of experience, of social order, of human relation."[7] In short, excavation projects were social metaphors. For the people of the nineteenth century, the everyday experience of looking at overturned soil provided a visual image of social upheaval.

The response of the middle classes was highly ambivalent. On the one hand, excavation projects were a supreme emblem of what the age understood as progress. As with scientific excavation, the enterprise itself was conceived of in heroic terms, as a bold foray into the depths of the earth, this time to achieve the physical rather than the intellectual domination of nature. But the excavation also aroused fear and anxiety, as it became clear that society as well as nature was being undermined. Established neighborhoods and communities were uprooted for the installation of railway or subway lines, and the rhythms of urban life were regularly interrupted by new interludes of digging. Paeans to progress alternated with complaints about these disruptions and with expressions of anxiety about the new order.

The story of this chapter is the construction of the new infrastructure, at once technological and social, of modern life. As such, it is also a story about the construction of consciousness. Excavation projects were metaphors for the undermining of an existing society, but they were also metaphors for the abstract progress of civilization. The conflict between the two responses to excavation projects—between seeing them as cruelly disruptive and seeing them as wonderfully heroic—represents the fundamental ambivalence of the middle classes toward the advent of a technological environment.

Since antiquity humans have dug into the earth to make tunnels for roads and aqueducts, to obtain useful minerals and rocks, and to wage war by undermining fortress walls or by providing escape routes for besieged towns. Some ancient subterranean projects were stupendous in scale. For example, the Romans dug two road tunnels, each over half a mile long, through the Ridge of Posilippo to connect Naples with Pozzuoli and an imperial villa.[8] By their nature, however, such projects of civil or military engineering were unusual and episodic. Mining too was set apart from ordinary life and labor, since it was usually performed in remote, isolated locations. In all cases, most of the subterranean laborers were serfs, slaves, criminals, or prisoners of war. The conditions they endured were indeed inhuman. The darkness was broken only by the feeble flicker of torches; the stuffy and often poisonous air was laden with choking smoke and dust; groundwater dripped unrelentingly; and there were the constant dangers of collapse and flooding, the constant need to stoop in the narrow, low passages, and, of course, the exhausting labor of shoveling and picking. Mining was a form of punishment. The social degradation of underground labor, as much as unconscious psychological terrors, explains why the underworld was dreaded as the region of sorrow and death.[9]

Only with the industrial revolution did underground enterprises became proximate, central, and large-scale, rather than remote, peripheral, and small-scale. Until then, wood and rocks had been the favored building materials and wind and water the favored sources of energy—all things that could be found on the earth's surface. But iron and coal, the fundamental material and energy source of early industrialization, had to be extracted from the earth. The industrial revolution brought an unprecedented increase in mining activity. In the 1720s, on his travels through the region around Newcastle, Daniel Defoe marveled at the "prodigious Heaps, I might say Mountains, of Coals, which are dug up at every Pit, and how many of those Pits there are." "We are filled," he wrote, "with Wonder to consider where the People should live that can consume them."[10] By 1750 Britain's annual coal output was 7 million tons. A century later, it would be 150 million tons.[11]

The crucial inventions of the industrial revolution also came from below the ground. The steam engine was first developed (in the form of the Newcomen atmospheric steam engine) to drain groundwater from mines. The motivation for James Watt's work on the steam engine was to improve the efficiency of the Newcomen device and also to develop an engine that could be used in manufacturing as well.[12] The first steam locomotive was built to haul ores. In the area around Newcastle, by the early 1800s a dense network of pithead railways covered the region between the mines and the river Tyne; by the 1820s, the horses that pulled the wagons were rapidly being replaced by steam locomotives, since coal was cheaper there than fodder.[13]

The industrial revolution implied a transportation revolution. If industry was to develop beyond very narrow limits, internal distribution of goods in bulk had to be improved. In Western Europe, where the terrain is quite flat and the rainfall abundant, water transport had always been much cheaper and more capacious than overland haulage. A natural system of internal waterways compensated for backward and inefficient overland transport. The first great step in improving this network, taken between the late 1600s and the early 1800s, was ambitious canal construction. Unlike natural rivers, which were limited by rapids, floods, or droughts, canals approximated a mechanical ideal, "striking across country, piercing the hills by tunnels, crossing valleys by embankments and rivers by aqueducts, and climbing to their summits by flights of locks."[14]

The shift from rivers to canals, from following the terrain to reshaping it, required feats of excavation. The French were the first to undertake large-scale canal construction. The Briare Canal, joining the Loire to the Seine, was completed in 1642, and between 1661 and 1681 the 148-mile Languedoc canal was built to link the Bay of Biscay with the Mediterranean Sea. The Languedoc Canal flowed through the Malpas Tunnel, 515 feet long and 22 feet wide, which had been blasted through solid rock with gunpowder. In his widely read history *Le Siècle de Louis XIV,* Voltaire mentions other monuments constructed by the Sun King, including the Louvre and Versailles, but concludes that "the most glorious, on account of its utility, its grandeur, and its difficulties, was the canal of Languedoc."[15]

When Francis Egerton (later the third Duke of Bridgewater) visited the Languedoc Canal in 1754, he was so impressed that upon his return to England he undertook the construction of a great canal from his mines at Worsley to Manchester. The Bridgewater Canal, opened in 1761, penetrated a mile underground into the mines, thus providing both easy transport and groundwater drainage. James Brindley, the civil engineer commissioned to build the Bridgewater Canal, went on to construct the Trent and Mersey (Grand Trunk) Canal, which required even more tunneling. Along its 93-mile length the Trent and Mersey Canal flowed through five tunnels, including, at its summit, the 9,000-foot-long Harecastle Tunnel, which had 35 locks to its west and 40 to its east. A prime supporter (and eventually the treasurer) of the Trent and Mersey Canal was Josiah Wedgewood, who valued it because it provided cheap transport to and from his potteries. Between the 1790s and 1830 about 3,000 miles of canals were built in England and Wales; the system eventually included 45 tunnels, totaling over 40 miles.[16]

The heroic age of transportation tunnels opened with the advent of railroad construction in the 1830s. In that decade the railroad ceased to be an appendage of the mine and evolved into the major distribution system of the nineteenth century. Like the canal, the railroad was conceptualized as a straight and level line. Traditional overland roads needed tunnels only in extraordinary circumstances (such as the need to pierce the Ridge of Posilippo), because it was assumed that roads, like rivers, were part of the terrain and would go around most natural obstacles. In the case of the railway, however, conquering nature meant leveling it. The function of the rail was not only to minimize the resistance caused by friction but also to smooth out nature's irregularities. The more perfect the track, the more it resembled a straight and level line—a geometric ideal imposed upon the bumps and hollows of the natural landscape.

Early railroad designers assumed the railroad would work only on a nearly level grade. Smooth steel on smooth steel did not seem capable of sufficient adhesion for uphill or downhill hauling. Eventually these doubts were banished by simple experience, but the concept of a roadway designed with ruler and level also had a certain philosophical appeal. Such a roadway conformed to the Newtonian mechanical ideal of a trajectory—perfectly hard, straight, smooth, and frictionless.[17] Another major attraction of the straight and level roadway was its cost. Because land was expensive and labor relatively cheap in the Old World, it made financial sense there to construct tunnels, embankments, and cuttings so the railroad would be as short as possible and the land costs minimized. Economic conditions in the United States were just the opposite. American railroad lines tended to be more curved and to go around natural obstacles rather than confronting them—not out of respect for nature, but because land was so much cheaper.[18]

In Britain and Europe the advent of the railway age meant that tunnels (as well as bridges and viaducts) were constructed on an unprecedented scale. In the late 1820s, when George Stephenson was asked to design the first regularly operating railroad line in England (to connect the port of Liverpool with the industrial center of Manchester), he planned a route with a grade of no more than 1 in 800, except for one short section with a grade of 1 in 96 where added tractive power was provided. Near Liverpool, Stephenson constructed the 6,750-foot Edge Hill Tunnel through soft clay, wet sand, and shale, all of which were extremely difficult to excavate. In 1836, Isambard Kingdom Brunel began constructing the Great Western railroad between Bristol and London. This famous line was just under 119 miles long and included the Box Tunnel (east of Bath), which was nearly 2 miles long, with a grade of 1 in 100.[19]

Even more striking victories over nature would come on the Continent, where even the mighty Alps could not stand in the way of the railroad. In 1857 work began on the first of the three great Alpine tunnels, this one stretching 8 miles under Mount Cenis to connect France and Italy. The Mt. Cenis Tunnel pierces a solid rock mountain at a maximum depth of 4,000 feet. The two sides met on Christmas Day 1870, and the tunnel opened to traffic three months later.[20] The next year, work began on the St. Gotthard Tunnel linking Germany and eastern Switzerland with Italy. The Simplon Tunnel, begun in 1896, was even longer (12¼ miles), deeper (a maximum of 7,005 feet), and technically more difficult than its predecessors, but it was completed in just 8 years. By the first decade of the twentieth century, a traveler could speed from Paris to Turin in under 18 hours.[21]

Such projects incarnated what the age understood as progress. Railway tunnels are prime examples of what art historian Kenneth

The tunnel under Edge Hill, on the Liverpool and Manchester Railway. From Thomas Talbot Bury, *Colored Views on the Liverpool and Manchester Railway* (London, 1831). Courtesy of Rotch Library, Massachusetts Institute of Technology.

The Moorish arch spanning the entrance to the Edge Hill tunnel. From Thomas Talbot Bury, *Colored Views on the Liverpool and Manchester Railway* (London, 1831). Courtesy of Rotch Library, Massachusetts Institute of Technology.

Clark has called "heroic materialism." Their construction was, in Clark's words, "like a great military campaign: the will, the courage, the ruthlessness, the unexpected defeats, the unforeseen victories."[22] The Edge Hill Tunnel on the Liverpool-Manchester line was decorated with a grandeur usually reserved for military victories: one end was flanked by two enormous chimney stacks decorated like Roman columns, while the other end emerged under a triumphal arch that resembled ones recently uncovered in Egypt. Lithographs of this and other famous tunnels were circulated throughout Britain. When John Cooke Bourne produced a series of prints of Brunel's Great Western Railway, the frontispiece set the tone for the collection. It showed a steam engine, its copper boiler gleaming, bursting full steam ahead from the gloom of a tunnel into the dazzling light of day. "Never before or since," writes the cultural historian Francis Klingender, "has anyone interpreted the simplicity, boldness and drama of great engineering with such deliberation and such verve."[23]

An equally heroic vision of subterranean engineering is evident in Charles Kingsley's novel *Alton Locke* (1850). In a dream sequence that presents a symbolic history of civilization, Kingsley tells how the community of the future performs two different kinds of labor: cultivating the land in a valley and boring through a mountain. Farming ensures the physical survival of the community, but tunneling expresses its spiritual mission; the mountain is the obstacle that stands between the human race and its western place of origin and destiny. When the community ceases work on the tunnel, it falls prey to class conflict. At that point, "Alton—who has remained faithful to work on the mountain—becomes their leader; when they follow him, the original state of equality and harmony is restored."[24]

Like the triumphal entrance to the Edge Hill tunnel, and like the lithographs of Bourne and other artists, Kingsley's novel invested tunneling with symbolic stature, making it a metaphor for the abstract progress of civilization. This glorification, though, contradicted the realities of subterranean labor. While excavation and tunneling were being glorified as emblems of civilization, the work itself remained brutal.

In Britain the heroes of excavation were the railway navvies. The finest of them came from the low-lying fen areas around Lincoln and Cambridge, where they had gained experience in the techniques of excavation and in the evaluation of soils and rocks. For the middle-class public, the navvies, who wore kerchiefs and heavy laced boots, seemed the cream of the working class—sturdy, independent, and diligent. In the words of Samuel Smiles in his *Lives of the Engineers,* "They displayed great pluck, and seemed to disregard peril."[25] In fact, though, peril was all too real. Except for the use of gunpowder to break up rock, excavation work had not changed much since antiquity: pickaxes, shovels, animals to haul away debris, flickering light, constant danger. As Smiles acknowledged, for the navvies "accidents [were] of constant occurrence."[26] To dig a deep embankment, for example, they had to hitch a wheelbarrow to a horse and then guide the barrow up a sloping board. If the horse's motions were at all irregular, the barrow, filled with soil and rocks, could come tumbling down on top of a worker if he failed to jump out of the way in time. Over a hundred navvies perished in building the Great Western railroad.[27]

Men working on the excavation of Olive Mount, 4 miles from Liverpool, on the Liverpool and Manchester Railway. From Thomas Talbot Bury, *Colored Views on the Liverpool and Manchester Railway* (London, 1831). Courtesy of Rotch Library, Massachusetts Institute of Technology.

Although mechanization did little to make excavation work safer or easier, it did much to speed up the work. The effects of mechanization on efficiency first became evident during the construction of the Hoosac Tunnel in western Massachusetts. When work began in 1851, the builders placed great hope in a 75-ton steam-driven drilling machine. A year later the machine broke down after progressing only 10 feet, and the workers returned to traditional methods. In 1856 another steam drill was tried, but it broke down before advancing even a foot. In 1866, a local mechanic named Charles Burleigh designed a small, light drill, powered by compressed air, that did prove workable. In 1868, the Hoosac workers began to make regular use of nitroglycerin, an explosive superior to gunpowder. As a result, the rate of advancement in the Hoosac Tunnel came to average 16–30 feet per face per week, whereas in previous American tunnels the average had been 9–16 feet per face per week.[28]

Similar improvements in efficiency were evident during the construction of the Mt. Cenis Tunnel, begun somewhat after the Hoosac Tunnel but also completed in the early 1870s. Power drills—most notably the pneumatic type devised by Germain Sommeiller, the engineer in charge of the Mt. Cenis project—were first used on the Italian heading in 1861 and came into use on the French heading two years later. Before then, with hand drills, work had averaged 12 feet per week per face; afterward, the average went up to 32 feet per week. By 1870 the combined yearly advance on the two headings was nearly 10 times what it had been when the tunnel had been started 13 years earlier. The St. Gotthard Tunnel, begun in 1872, was even longer (9⅓ miles) and deeper (nearly 6,000 feet at its maximum) than that through Mt. Cenis; however, it was completed in only 9 years rather than 14, thanks to even better drills and to the use of dynamite.

Tunneling remained killing work, however. In the 9 years required to dig the St. Gotthard Tunnel, 310 workers died and nearly three times that number were rendered invalids by diseases and accidents. Technical advances did not necessarily translate into safety. Besides facing the dangers of explosions and falling rocks, the workers were often disabled by silicosis, pneumonia, and other lung diseases caused by inhalation of rock dust.[29]

These working conditions, however, were not widely known; like the tunnels themselves, the workers were hidden from the daylight of public awareness. The middle-class public was being introduced to the perils of subterranean environments in other ways, however. Through most of history, those environments had been experienced only by those at the bottom of the social ladder—convicts, prisoners of war, slaves. In the nineteenth century, for the first time, railway tunnels opened up the underground experience to those on the middle and the top rungs of the ladder. The experience of disconnection from nature, of immersion in a manufactured environment, was no longer restricted to social outcasts. Any railroad passenger experienced the loss of contact with the landscape, and most notably so when going through a tunnel.[30]

For the traveler as for the miner, disconnection from nature meant peril. Train accidents were common enough anyway; and with no light and with no escape route, tunnels posed particular dangers. On early trains the basic safety device was the human observer, who, mounted on the carriages, would scan the track ahead for dangerous obstacles or oncoming trains. In tunnels, where human beings are detached from the landscape, this organically based system would not work. Here sensory observation was replaced by mechanical systems—various optical and auditory signals at first, then the far more effective telegraph signals. In fact, tunnel passages provided the first practical application of the telegraph. Beginning in the 1830s, first in tunnels and then all along the line, the "space interval" or "block" system was introduced: the railroad line was divided into separate blocks, each regulated by a telegraph transmitter which signaled ahead when the line was clear.[31]

In the pre-industrial era, the concept of accident was primarily "grammatical and philosophical" and "more or less synonymous with coincidence."[32] By the middle of the nineteenth century, however, the term *accident* was used almost exclusively to refer to technological mishaps, especially those involving railways. The number of railway passengers injured or killed was, of course, trifling in comparison with the casualties suffered by the workers who had constructed the cuts, bridges, and tunnels that furnished a level roadbed. In cultural terms, though, railway accidents were unprecedented, for their victims were more likely to be persons of wealth and status. In the form of accidents, technological progress brought a certain democratization of disaster. From the middle-class perspective, there seemed an inevitable and necessary relation between technological advance and human sacrifice. It is not just that progress was seen as having its price, but that progress seemed to *demand* its price; accidents were the sacrificial offering to appease that god. In this way too, a remnant of the mythological stalked the corridors of the nineteenth century.

Railway accidents were one example of the dialectical relation of progress and destruction; railway construction was another. In this case not bodies but communities were mangled. When tracks were laid in populated lowland areas, the goal of conquering nature inevitably meant reshaping not just landforms but communities too. Human dwellings, like hills, were obstacles in the way of the straight and level roadbed. In 1836, for example, when the London and Birmingham Railway was laid down, its builders cut a swath from Euston through the streets and tenements of Camden Town and then cut a tunnel through Primrose Hill, from whence the track proceeded north to Birmingham. In the novel *Dombey and Son*, published serially from 1846 to 1848, Charles Dickens describes the destruction of Camden Town (where a nurse is found for the motherless infant Paul Dombey) as if it had been overturned by a natural disaster:

The first shock of a great earthquake had, just at that period, rent the whole neighborhood to its center. . . . Houses were knocked down; streets broken through and stopped; deep pits and trenches dug in the ground; enormous heaps of earth and clay thrown up; buildings that were undermined and shaking, propped by great beams of wood. Here, a chaos of carts, overthrown and jumbled together, lay topsy-turvy at the bottom of a steep unnatural hill. . . .

In short, the yet unfinished and unopened Railroad was in progress; and, from the very core of all this dire disorder, trailed smoothly away, upon its mighty course of civilization and improvement.[33]

Dickens is being ironic here, but not completely so. He never resolves his ambivalence. He fears the upheaval caused by the railroad, yet he welcomes it as an improvement. As the Marxist critic Terry Eagleton notes, *Dombey* is split "between a conventional bourgeois admiration of industrial progress and a petty-bourgeois anxiety about its inevitably disruptive effects."[34] Dickens emphasizes the disruption in the section of the novel where the dying Paul Dombey asks to see his old nurse again. A family friend goes to search for her in Stagg's Gardens, the part of Camden Town where she had lived, and discovers

There was no such place as Stagg's Gardens. It had vanished from the earth. Where the old rotten summer-houses once had stood, palaces now reared their heads, and granite columns of gigantic girth opened a vista to the railway world beyond. . . . The old by-streets now swarmed with passengers and vehicles of every kind: the new streets that had stopped disheartened in the mud and wagon-ruts, formed towns within themselves, originating wholesome comforts and conveniences belonging to themselves, and never tried nor thought of until they sprung into existence.[35]

Dickens goes on to enumerate the benefits of progress—the "throbbing currents" of movement, both of people and of goods, that had meant a new house for the family and a new job for Mr. Toodle on the railroad, safer employment than his former work in the mines. Still, Dickens concludes the description with a lament: "But Stagg's Gardens had been cut up root and branch. Oh woe the day when 'not a rood of English ground'—laid out in Stagg's Gardens—is secure!"[36]

What Dickens is confronting here, of course, is the dialectic of progress and destruction: building a tunnel also means undermining a neighborhood. The dominant cultural response, from Dickens's time to our own, is a reasoned debate about the price of progress— an abstract weighing of costs versus benefits, an analysis of comparative risks, an attempt to balance the social ledger. The debate is endless because the real issues are not the abstractions, presented as absolutes, but the relative experiences of particular individuals in particular situations. Someone who holds railway bonds will likely evaluate the situation quite differently than someone in Stagg's Gardens whose house has been leveled. Dickens is so ambivalent because he is part of the middle class in more ways than one. He identifies both upward and downward. He praises the gains in national wealth and convenience, but because of his imaginative powers and his own experiences with poverty he also empathizes with those whose neighborhood has been destroyed.

Effective political action is not based on deep ambivalence. The debate about the price of progress continued, but it never interfered with the building of railway tunnels. It was in the cultural, not the political realm, that the anxieties about industrialization were primarily expressed. And in the cultural realm, those anxieties were repeatedly shaped into subterranean images that were simultaneously used as emblems of progress.

Once again we pivot from technology-as-fact to technology-as-metaphor. As we have seen, for Lewis Mumford the industrial revolution was both factually and metaphorically an eruption from the depths, a victory of the underworld that transformed the green world of the surface into the black world of the mine. But Mumford was not the only one who, in retrospect, characterized paleotechnic industry as an eruption from below. From the early eighteenth century on, British travel literature overflows with descriptions by middle-class people who had visited mining areas and had been stunned by the black pits, the blighted landscape, the smoke and steam, the strange noises of machinery, the flaring lights of furnaces, and the dark, demonic human figures. These associations between industry and hell grew even stronger in the early nineteenth century, which Klingender and many others call "the age of despair." By then the Enlightenment's bright hopes for a smooth and seamless advancement of science, technology, and society had been disappointed by decades of political and economic dislocation. In the summer of 1813 Richard Ayton visited the William Pitt mine, reputed to be one of the newest and best-planned in the British Isles, and reported: "A dreariness pervaded the place which struck upon my heart—one felt as if beyond the bounds allotted to man or any living being, and transported to some hideous region unblest by every charm that cheers and adorns the habitable world. . . . All the people whom we met . . . looked like a race fallen from the common rank of men, and doomed, as in a kind of purgatory, to wear away their lives in these dismal shades. . . ."[37]

The metaphor worked in the other direction, too: hell was given the image of industry. In the 1820s and the 1830s, John Martin, besides illustrating Mantell's *Wonders of Geology,* illustrated *Paradise Lost* and the Old Testament with images drawn from tunnels and mines, giving Satan's realm the forms of industry.[38] In Martin's mezzotints for *Paradise Lost,* Pandemonium (the City of Hell) resembles the stepped landscape of a Welsh slate quarry, and the details of the city itself echo images of mines, of the underground vaults below

The Bridge over Chaos. Mezzotint illustration by John Martin for Milton's *Paradise Lost* (1827). Courtesy of the Board of Trustees of the Victoria and Albert Museum.

Pandemonium, Satan's city in the underworld. Mezzotint illustration by John Martin for Milton's *Paradise Lost* (1827). Courtesy of the Board of Trustees of the Victoria and Albert Museum.

the London docks, and of canal tunnels. The mezzotint showing Sin and Death building a bridge over Chaos shows a viaduct within an enormous hollow tunnel—within nothingness.[39]

In tracing the connection between industrial reality and industrial metaphor, the core image is the mine, the literal underworld. Other industrial locales, not actually underground, mimicked subterranean conditions so faithfully that they too were compared to hell. William Blake's "dark Satanic mills" were not actually buried structures, but they might as well have been. Though the earliest cotton mills had classical proportions and resembled country houses, their successors came to resemble fortresses. Like prisons, they were enclosed by forbidding walls, so that the workers were shielded from any distractions.[40] But the circles of metaphor rippled even further. Not just mines, not just factories, but whole industrial districts seemed to resemble eruptions from the underground, to be regions where nature had disappeared. This is how a traveler described the "Black Country" of England in 1813:

From Birmingham to Wolverhampton, a distance of thirteen miles, . . . part of [the country] seemed a sort of pandemonium on earth—a region of smoke and fire filling the whole area between earth and heaven; amongst which certain figures of human shape—if shape they had—were seen occasionally to glide from one cauldron of curling-flame to another.[41]

Three decades later, Charles Dickens used the same area around Wolverhampton as the setting for the final tribulation of Little Nell and her grandfather in *The Old Curiosity Shop*. He too described the landscape as hell on earth:

. . . they came by slow degrees upon a cheerless region, where not a blade of grass was seen to grow. . . .

. . . to the right and left, was the same interminable perspective of brick towers, never ceasing in their black vomit, blasting all things living or inanimate, shutting out the face of day, and closing in on all these horrors with a dense dark cloud.

But night-time in this dreadful spot!—night, when the smoke changed to fire, when every chimney spurted up its flame; and places, that had been dark vaults all day, now shone red-hot, with figures moving to and fro within their blazing jaws. . . .[42]

In this lurid and frightened description, Dickens was responding to a particular historical moment: the Chartist uprisings of the 1840s. However, the same associations—coal and iron, darkness and flames, demons and depths—were repeated over and over by other observers at other stages of the industrial revolution.[43]

Even in the twentieth century, in some regions of heavy industry, these associations still endured. Nearly a century after Dickens wrote, George Orwell, following the road to Wigan Pier, stayed in a lodging house in a mining area which he too described as an underworld—but as a mean little underworld, expressive of the despairing apathy of the Great Depression. He decided to leave his lodgings, Orwell says, when he became unbearably depressed by "the feeling of stagnant meaningless decay, of having got down into some subterranean place where people go creeping round and round, just like blackbeetles, in an endless muddle of slovened jobs and mean grievances."[44] The departing train bore Orwell "through the monstrous scenery of slag-heaps, chimneys, piled scrap-iron, foul canals, paths of cindery mud. . . ." He was amazed when the train pulled into open country. It seemed "strange, almost unnatural," for "in a crowded, dirty little country like ours . . . slag-heaps and chimneys seem a more normal, probable landscape than grass and trees, and even in the depths of the country when you drive your fork into the ground you half expect to lever up a broken bottle or rusty can."[45] The train rolled on for another 20 minutes or so through a strange and lovely landscape of snow, birds, and bright sunshine. Then "villa-civilization began to close in upon us again," and another industrial town banished nature once more.

If the industrial landscape, from the time of Dickens to that of Orwell, seems to have spread subterranean evil over the earth's surface, the *products* of industry seem to have spread heavenly benefits. Orwell's lodging house, for all its meanness, had electricity, a kitchen range, and cold running water. Even in the poor areas of Britain in the 1930s, the towns had movie theaters, and many people could listen to phonographs and radios or buy inexpensive books. According to Orwell himself, what most reminded him that "our age has not been altogether a bad one to live in" was "the memory of working-class interiors": the winter evenings when, after kippers and strong tea, Father would sit in a comfortable chair on one side of the coal fire to read the paper, while Mother would do her sewing on the other side as the children nibbled candy and the dog lolled on the rug.[46]

Many of the comforts Orwell mentions are products of the second industrial revolution of the late nineteenth century. The distinctive artifacts of the first industrial revolution had been the machines of production, especially those related to metalworking and textile manufacture. The distinctive artifacts of the second industrial revolution were consumer goods such as the bicycle, the automobile, chemical dyes, the telephone, electric lights, the camera, the phonograph. Never before nor since has there been such a concentrated period of technological change affecting ordinary people. What they ate, what they ate with, where they lived, what they wore, how they traveled, how they amused themselves—all these daily items and activities were being altered simultaneously. This was not progress on a heroic, titanic scale; it was domestic, even intimate progress, which brought health and convenience and comfort to daily life.

If the creation of a network of canals and railroads had been the first stage in the creation of a modern industrial infrastructure, the creation of urban networks defined the second stage. For the first time, the city was thought of as a unified system, its proper functioning depending on buried subsystems—networks of electrical power, of sewer and water service, of subway and telephone lines. In the networked city, the integrity of neighborhoods was less important than the functioning of the metropolis as a whole. Only when the underground was crowded with utility lines did it become possible to provide relatively decent living conditions for so many people crowded together above the surface. The networked city made possible the city of skyscrapers.

Water and sewer systems were usually the first networks to be constructed. In most cases those systems had long been in place, but in the nineteenth century they were vastly expanded and improved. Victor Hugo was so impressed by the scale and difficulty of the renovation of the Paris sewer system that he devoted several long and detailed chapters of *Les Misérables* to it. At the beginning of the century, as Hugo recounts, the Parisian sewer system was basically unchanged from that of the seventeenth century. In 1830 the city had about 40 miles of sewers. Another 60 miles were built between 1830 and 1850. By then, however, it was clear that the growth of the city had outstripped the capacity of the system. The Seine had become an open cesspool. Furthermore, because Paris is built on such low land, any flood interrupted sewer service and backed up wastewater, which flooded basements and streets and which caused the streets to be covered with ice in the winter. In terms Kenneth Clark would appreciate, Hugo compared the work to a military campaign:

Paris is built upon a deposit singularly rebellious to the spade, the hoe, the drill, to human control. . . . The pick advances laboriously into these calcareous strata alternating with seams of very fine clay and laminar schistose beds. . . . Sometimes a brook suddenly throws down an arch which has been commenced, and inundates the labourers; or a slide of marl loosens and rushes down with the fury of a cataract, crushing the largest of the sustaining timbers like glass. . . . After having built the Saint Georges sewer upon stone-work and concrete in the quicksand; after having directed the dangerous lowering of the floor of the Notre Dame de Nazareth branch, Engineer Duleau died. There are no bulletins for these acts of bravery, more profitable, however, than the stupid slaughter of the battle-field.[47]

During the Second Empire, from 1852 to 1870, underground Paris was further transformed. Baron Georges-Eugène Haussmann, Prefect of the Seine under Napoleon III, is best known for rebuilding Paris above the ground. As Schivelbusch writes, "In a mere fifteen years the physiognomy of that city underwent a complete transformation, a 'regularisation' (Haussmann) that is unique in European history."[48] At the same time, though, Haussmann and his chief of water services, Eugène Belgrand, regularized subterranean Paris. They gave nearly every street an underground drain and built a comprehensive system of large collectors to carry wastewater into the Seine well below Paris. In the late 1850s Belgrand constructed a huge collector sewer, which began at the Place de la Concorde and eventually discharged its effluent into the Seine 20 miles to the north. To build such a deep conduit in unstable, water-bearing soils was a remarkable feat. Under Belgrand's direction, the hideous labyrinth that Jean Valjean supposedly escaped through in 1832 was transformed into (in Hugo's words) a "neat, cold, straight, correct" sewer. It was far more comprehensive as well. When Haussmann and Belgrand began, Paris had under 100 miles of sewers; by 1870, at the end of the Second Empire, the city had 348 miles, or nearly four

A subterranean street scene in Paris. From Edmond Texier, *Tableau de Paris*, volume 1 (Paris, 1852). Courtesy of Widener Library, Harvard University.

times the total of 1851.[49] Thus, as Hugo notes with pride, the Parisian sewer system had expanded tenfold between 1800 and 1870.

Once the work was completed, Paris had a system of multipurpose underground galleries. In sewers built there after 1855, the actual sewer runs in a trough at the bottom of a large, egg-shaped tunnel made from hydraulic cement. The galleries were large enough so that other networks could be suspended from the walls and roofs—water lines, gas lines, steam pipes, pneumatic tubes, compressed air lines, and cables for telephone and telegraph and electricity. The Parisian system of underground galleries, which had no equal anywhere in the world, quickly became a tourist attraction. During the international exposition of 1867 both visiting royalty and common tourists inspected *les égouts de Paris*. Baedeker recommended them, and special facilities had to be set up to handle the sightseers.[50]

Similar work went on in London, where the drainage system was overhauled between 1859 and 1865. A new grid of sewer lines was constructed to intercept the contents of the old ones and carry the contents not directly to the Thames (which had become a stinking

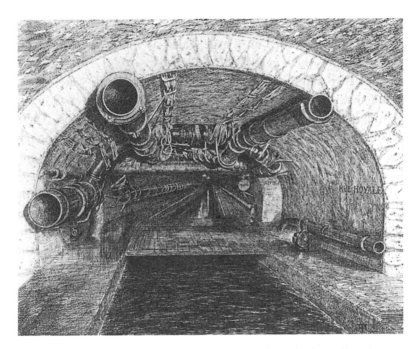

The vault of a large collecting sewer under the rue Royale. From Paul Strauss, *Paris ignoré* (Paris, 1892). Courtesy of Widener Library, Harvard University.

breeder of cholera epidemics) but to an outfall 14 miles below London Bridge. The new sewers, laid perpendicular to and somewhat lower than the existing ones, required massive excavations: 82 miles of intercepting sewers were dug, and 1,300 miles of ordinary ones. Tunnels as deep as 60 feet were dug under canals, rivers, railways, and residential areas. One house adjoining a railway station was underpinned and placed on iron girders while a sewer more than 9 feet in diameter was carried through the cellar "without further injury to the house."[51]

In sheer scale, though, the most monumental projects were the subways. Approximately 1,070,000 cubic meters of rock and soil were removed for the Simplon Tunnel through the Alps; for the London tubes, 4,500,000 cubic meters.[52] Furthermore, the construction of extensive subway systems demanded considerable technological innovation. In most cities other than London—Paris, New York, and Boston, for example—the early subways were only 10–15 feet below the surface. At such depths the cheapest methods were "cut-and-cover" techniques similar to those used to lay sewer and water lines. The tools were simple (spades, pickaxes, crowbars, and the like), but when used on a large scale—as was possible in an age of cheap labor—they were highly effective.

The first step in cutting (whether for sewers, for water pipes, or for shallow subways) was to loosen the soil with a plow—a reminder of the countryside in the midst of the metropolis. Then, men using picks would hack at the soil while others wielding shovels would throw the dirt onto scrapers (dragged or wheeled), onto wheelbarrows, or, more often, onto dump cars drawn by horses along a light iron rail. Sometimes steam shovels were used to dump the soil directly into cars or carts, which were then hauled away by animals or by locomotives. In a prose poem published in 1911, the French author Jules Romains describes the heavy rhythm of a gang working on the Métro:

The pickaxes keep nothing of what they bite; they are not filled. Their mouthfuls pile up at the feet of the men and bury their shoes. But the shovels lap up clay and throw it into the tip-truck that a horse with hairy hooves is going to pull on the muddy rails. . . .

When the tip-truck is filled, the horse stiffens its haunches; the load rises, and the axles chirp. . . .

Street scene in Paris during repair work on a sewer. From Paul Strauss, *Paris ignoré* (Paris, 1892). Courtesy of Widener Library, Harvard University.

Then the gang turns the crank, lengthens the hooks, takes the tip-truck by
the armpits, lifts it up, balances it at the end of its chain, and tips it on the
cart where clods tumble down, muttering. . . .[53]

In rocky ground, where holes had to be drilled and explosives used, the workers used special precautions to prevent harm to buildings and bystanders. Areas where charges were placed were covered by logs chained together or by a rope mat weighted with logs to keep the loosened rocks from scattering too far.[54] Once the hole was made, side retaining walls would be built (from iron ribs and brick, or later from concrete and steel beams), and then a roof; the street would be restored afterward.

Any extensive system of urban transportation, however, required deeper tunnels and more complicated technology. In Paris, where the first Métro line opened in 1900, conventional tunneling had to be used at some points because of the slope of the ground. For example, one line between Montmartre and the Gare Montparnasse has an average depth of 40 feet but goes down to 114 feet at the Place des Abesses station. In London too, expanding the subway system required deep tunneling. As early as 1860 a relatively shallow "inner circle" of subway track had been started to connect twelve London railroad stations, but enlarging the system to encompass the entire city required tunnels that would pass under the Thames.

In such cases the major technical problem was how to tunnel through water-bearing soils. Wet ground, springs, and long-buried streams always presented difficulties in Paris, where the geology is particularly varied and complicated. London and other cities built along rivers presented similar problems of unstable earth laced with groundwater. The headings and timberings developed for conventional mines would not work in the treacherous underworld of quicksands, oozes, mucks, and muds, where water exerted tremendous pressure and always threatened to flood the excavations.

The first person to make a sustained attempt to conquer the underworld of mud was Marc Isambard Brunel (1769–1849), a French-born engineer who emigrated to England in 1793 and who was the father of Isambard Kingdom Brunel. In 1826 the Brunels began constructing a road tunnel under the Thames east of the Tower, between Rothertithe and Wapping and 3 miles east of Charing Cross. The son, barely out of his teens, was the engineer in charge.

Brunel *père* had designed and built an enormous rectangular shield consisting of twelve vertical cast-iron frames, each 22 feet high and 3 feet wide and each with three vertically stacked compartments. A shoe at the bottom of the shield was pushed forward by a large screw, and workers occupying the compartment bricked up the tunnel walls as the shield was pushed forward. The mud they excavated was carried to wheelbarrows and then wheeled to a shaft, where it was raised in buckets by a windlass driven by a steam engine, which also ran the drainage pumps. The method was excruciatingly slow (maximum daily progress was 10 inches) and terribly dangerous. The top of the tunnel was only 16 feet below the bed of the Thames, and water flooded in numerous times, sometimes drowning the workers. In order to reassure the public and the shareholders, the Brunels held a banquet in the tunnel on the evening of November 10, 1827. The tunnel walls were covered with crimson cloth, chandeliers fueled by portable gas burned in decorative urns, and the Band of the Coldstream Guards played while the workers (who has been invited along with the shareholders) toasted their tools and presented the Brunels with a pickaxe and a spade symbolizing their labor. A few weeks later the tunnel was flooded yet again. Work ceased until 1835, when more funds were raised, and Marc Brunel then went back to work with a redesigned shield. (By that time, Isambard Kingdom Brunel was at work on the Great Western.) The 1,200-foot tunnel, completed in 1841, opened in 1843 with great fanfare, including commemorative prints, medals, and handkerchiefs.[55]

No extensive system of underground transportation was possible until tunneling through soggy soils could be done much more rapidly and safely. Not until 1869 was another Thames tunnel dug, but this one took just 11 months to finish.[56] The much-improved shield used in that tunnel was designed by two engineers on the project, Peter William Barlow and his 25-year-old assistant James Henry Greathead. The Greathead shield, as it came to be known, was much smaller and lighter than Brunel's; it was only 8 feet in diameter and 3 feet long, and it was circular rather than rectangular. As it was forced ahead by hydraulic rams (it had a sharpened circular ring in front to help it drill soil), the shield was built up from behind by narrow iron rings bolted together to form a segmented cast-iron lining for the tunnel, rather than the brickwork Brunel had used.

When the Greathead shield was combined with compressed air that resisted water intrusion, deep tunneling through water-bearing soils became practical.[57] In 1886, just two years after the "inner circle" was completed, the first deep tunnels of the London tube system were begun. (The tubes are in fact cylindrical tunnels measuring 12 feet in diameter between stations and 21 feet in diameter at the stations.) The first tube line, the 7½-mile-long City and South London Railway line, was constructed by Greathead, who used his shield and compressed air to complete the line in just 4 years without a single fatality. Succeeding tubes were sunk at an average depth of 50 feet, but sometimes they went far deeper (the Hampstead Heath station lies 291 feet below the surface). New lines, all built with shields and compressed air, continued to be built deep in the London clay, gravel, and sand up to 1907.[58]

Compressed air was also used to lay the foundations of the tall, heavy buildings and bridges that began to be built in the later 1800s. The types of foundations that had served for centuries—continuous stone or masonry in solid ground, or wooden piles sunk into wet and unstable soils—were not adequate for much heavier structures. In this case, technological innovation moved from the New World to the Old. In Chicago, settling was a particular problem because the Loop area is underlain by wet clay laced with pockets of sand and water. Chicago builders would commonly lay sidewalks canted upward at an angle from the curb line, in hopes that when the adjacent building settled the sidewalks would sink with it to a level plane.[59] After the Great Fire of 1871, architects experimented with a number of new foundation types, including a "floating raft" of steel rails and long-pile foundations driven down to the bedrock, which was typically 75 or 100 feet below the surface. In planning the Stock Exchange building in 1894, the architectural firm of Adler and Sullivan devised a new solution: the sinking of massive concrete piers made from deep watertight drums that served both as forms for the concrete and as protection against water seepage. These open caissons, which became known as "Chicago caissons," were to serve as reliable foundations for many tall buildings, on both sides of the Atlantic.[60]

The excavation of foundations for some buildings and bridges required pneumatic caissons, open at the bottom but closed above and filled with compressed air to keep out water. As early as 1851, Isambard Kingdom Brunel sank a pneumatic caisson more than 70 feet long in the process of building his last great railway bridge, the Royal Albert in Cornwall. By the 1860s the technique was generally used in Europe and America. In 1870 a caisson more than 100 feet deep was excavated to lay the foundations of the St. Louis Bridge. Even more impressive was the pneumatic caisson used for the Brooklyn Bridge (1869–1883), the towers of which were for many years the tallest (276 feet) in New York. The Brooklyn caisson, launched in 1870 and sunk in place in 1871, was a gigantic box 168 feet long and 102 feet wide, weighing 3,000 tons and covering more than half the area of a city block. Washington Roebling, chief engineer of the project, published a long account of the sinking of the caisson in the Brooklyn *Eagle* just about the same time that the English translation of Jules Verne's *Twenty Thousand Leagues Under the Sea* appeared, prompting the *Eagle* to editorialize: "The adventures of Colonel Roebling and his twenty-five hundred men under the bed of the East River are as readable, as he tells them, as any story of romance which has issued from the imagination of the novelist."[61]

Adventures, surely; but not romantic ones. Once again, the impulse to mythologize subterranean engineering as an emblem of progress was contradicted by the terrible realities of subterranean labor. Underground work has always been dangerous, and working under muck was exceptionally so. Workers descended into a stifling black hole, knowing they could at any moment be drowned in mire. When pressurized air began to be used, workers confronted new perils. Even under the best of conditions, the high pressures were debilitating; laborers on the St. Louis Bridge, who were subjected to pressures of 45 pounds per square inch at over 100 feet below the surface, could work only half-hour shifts. Worse yet, after their shifts they took only a few minutes to decompress. Fourteen workers died and many more were seriously afflicted by the excruciatingly painful and debilitating "bends," or "caisson disease." The nature of this affliction was so little understood that it was thought to have something to do with galvanic action, and the men were issued supposedly preventive bands made of alternating scales of zinc and silver to wear around the waist, the ankles, the arms, and the wrists. Although the French physiologist Paul Bert discovered the cause of the disease by the 1860s, and although the first "medical lock" to provide slow decompression was built by the contractors of the first Hudson River tunnel (1874–1908), it was not until the early twentieth century that the dangers of caisson disease were fully understood and preventive measures required.[62]

For most of the public, however, this danger too was hidden from view. Once again, the price of progress typically experienced by the middle classes was disruption rather than disease or death. The long-term purpose of building bridges and subways was to relieve surface congestion, but the short-term effect was the opposite. Cut-and-cover techniques, though the least expensive, were the most disruptive because they required tearing up an entire street until work was completed. When the first Métro lines were dug in Paris, the chief engineer, Fulgience Bienvenue, realized that interference with surface traffic would be minimized if the subway were to be built from the top down. Accordingly, he devised a new technique, a hybrid of cut-and-cover and tunneling. A gallery was first cut along the upper part of the tunnel. From this gallery were cut shafts to the street, through which dirt could be removed as the gallery was widened. Next, the tunnel roof was constructed with stone blocks. Trenches were dug down and out so that masonry walls, resting on concrete, could be built. Finally, a concrete tunnel floor was laid. The whole street did not have to be torn up, and traffic could continue to circulate while work proceeded below.[63]

Even so, throughout the Belle Époque, the magazines and news-papers and cafés of Paris overflowed with complaints about the disruption caused by the building of the Métro. Jules Romains began his multi-volume novel *Les Hommes de bonne volonté* [Men of Good Will], which was intended to give a panorama of prewar French society, by describing rush hour in Paris on a typical day in October 1908 as people tried to get to work through the mess caused by Métro construction:

The scaffoldings of the subway, which rose up all over the place like fortresses of clay and planks, armed with batteries of cranes, had ended by strangling the streets, blocking all the intersections. Not to speak of the fact that this driving of tunnels was undermining the ground in all directions and threatening Paris with collapse. (That very October 3rd, part of the parade-ground of the Cité barracks had fallen into the Châtelet-Porte d'Orléans subway under construction, and a mounted policeman's horse had suddenly been swallowed up by the abyss.)[64]

The everyday experience of looking down into "the abyss"—like the experience of taking a railroad journey for an earlier generation—was decisive in shaping technological and social consciousness. Even the most fundamental underpinnings of the social order, soldiers and policemen, could be swallowed up by this social *bouleversement*. Yet despite the occasionally ominous and always disruptive nature of urban excavation, this second epoch of subterranean construction seemed far less threatening than the first. Whereas mineshafts and railway tunnels were technologies of steam and flames and smoke, the water mains and electrical cables laid in the later nineteenth century brought light and convenience and cleanliness. Even Lewis Mumford, who condemned paleotechnics as an eruption from hell, welcomed the new wave of technology, which he called *neotechnics*. Mumford prophesied that neotechnics would be based on conservation rather than destruction, and on efficiency and order rather than chaos. "Light shines on every part of the neotechnic world," he wrote. "The dark blind world of the machine, the miner's world, began to disappear. . . ."[65] To Mumford and many others, the subterranean projects of the neotechnic age promised a manufactured environment more like heaven than hell.

Surveying the damage to the Thames Tunnel shield after the first serious episode of flooding. From Richard Beamish, *Memoir of the Life of Marc Isambard Brunel*, second edition (London, 1862). Courtesy of Widener Library, Harvard University.

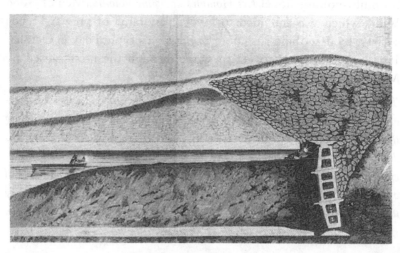

Between the late 1700s and the early 1900s, the ground of Britain and Europe was dug up to lay the foundations of a new society. Subterranean images became familiar sights during that period: workers sinking picks into the soil, city streets slashed down the middle, whole industrial regions turned into minelike terrains. These were, simultaneously, images of social and technological upheaval. In the cultural language of imagery, the middle classes began to establish connections between the new technological infrastructure and the new structures of society and consciousness. Subterranean iconography connects the historical experience of excavation and the literary interpretation of underworlds as technological environments.

Underground Aesthetics: From Sublimity to Fantasy

For one last time, let us call upon Lewis Mumford to set the stage. According to his aesthetic principles, the mine is irredeemably ugly, the opposite of beautiful. For Mumford, the fundamental principle of beauty—indeed, of all value—is organicism. Beauty results when form unfolds as naturally as a plant or animal grows, from the inside out. Such organic beauty can never be found in the underground environment, which is devoid of life and form, "a shapeless world: the leaden landscape of a perpetual winter."[1]

In stressing the ugliness of the underworld, Mumford follows an ancient aesthetic tradition. The mythic journey to the underworld is fearsome because the traveler descends to a realm of dampness, darkness, and formlessness. Bears, snakes, and even more frightful monsters (such as the Horibs of Pellucidar) are said to lurk there.[2] To work in a mine or a tunnel is to be condemned to pain and danger in a black, foul, cramped hole. Victor Hugo evokes all these associations when in *Les Misérables* he describes Jean Valjean's descent into the sewers of Paris, with Marius on his back: the dripping walls and low ceiling, the miasmas and pitfalls, the fetid odor, the obscurity, the heavy burden, the ominous shadows of criminals and policemen. The climax of the horror comes when Valjean steps into a fontis (a crevice of mud) and sinks up to his armpits, nearly drowning in slime and excrement.

It was Hugo, though, who urged historians to confront such ugliness in order to explore subterranean society. While it may make them shudder, he said, the social underworld is fully as significant as the upper crust. In a similar spirit, other artists and critics have sought aesthetic significance in the underworld since the later 1700s. In the first stage, they applied a new aesthetic principle, the sublime, to the dark world of the miner. The sublime underworld was neither ugly nor beautiful, but something else entirely: obscure but pleasingly obscure, terrible but delightfully so. Gradually, however, the underworld came to be perceived as wholly beautiful—as a magically illuminated artificial paradise, a splendid refuge from nature's imperfections. There are many exceptions to this general trend (Lewis Mumford, after all, was writing in the twentieth century), but the central evolution in the aesthetic of the underworld between 1700 and 1900 is from ugliness to sublimity to magical beauty.

This evolution in aesthetic values is closely related to technological changes, especially the transition from the age of steam to the age of electricity. The key difference between the sublime and the magical underworld is in the quality of illumination: the first is characterized by nocturnal flames licking up from below, the second by a mysterious glow emanating from a multitude of unseen sources. These dominant images of illumination were strongly suggested by contemporary technological practices. The image of licking flames evokes the fires and flares of paleotechnic industry (to borrow Mumford's term); the mysterious glow is characteristic of the neotechnic age of electricity.

If contemporary technological practices helped shape the aesthetic discourse, that discourse in turn shaped responses to technology. Aesthetic vocabularies first applied to the subterranean environment in particular were extended to the manufactured environment in general. Machinery too is not considered beautiful by traditional aesthetic standards, and may even be considered positively ugly. Like the subterranean environment, the mechanical one lacks aesthetic value when judged by principles based on organicism. Sublimity and fantasy, however, appeal to the superorganic. They convey the wonder aroused by phenomena beyond ordinary experience. The vocabularies of sublimity and magic that revealed superorganic value in the underworld also expanded the canon of beauty to include technological environments.

This expansion is much more than a chapter in the history of aesthetics. The history of art and taste holds many examples of "the transvaluation of what had previously been accounted ugly," and such transvaluations are "characteristically credited with being an enlargement of sympathy and a refinement of discrimination."[3] Ways of seeing are ways of valuing. As we shall see in this chapter, an aesthetic evaluation of the manufactured environment is also a social and moral judgment.

Before the eighteenth century, aesthetic inquiry was clearly understood to be part of metaphysical and moral philosophy. The central aesthetic problem was to define objective properties that make an object beautiful. In the "Copernician revolution"[4] of eighteenth-century aesthetics, however, the focus of inquiry shifted from the object to the observer. The central problem (as defined in that century, primarily by British thinkers) became one of feeling—of the psychological effect of beauty on the viewer, as opposed to the formal qualities of the object being viewed.[5] Once beauty was reconceptualized in this way, it seemed obvious there could be many types of aesthetic experience. Among the many varieties of aesthetic experience defined and described by eighteenth-century thinkers, the most important is sublimity.[6]

As a rhetorical style, sublimity had been defined centuries before in a treatise by the Greek thinker Longinus. Longinus's essay *On the Sublime* [*Peri Hypsous*] was rediscovered and translated in the sixteenth century, neglected during most of the seventeenth, and then revived in the eighteenth in a French translation by Boileau (*Traité de l'art poétique,* published in 1674). This translation, and numerous English ones that followed, made Longinus's work enormously popular in eighteenth-century England. (It was an anonymous translator of 1698 who first translated *hypsos* as *sublime,* a term already in use among critics.[7]) Joseph Addison—who has much as anyone can be called the father of modern aesthetics—was the first to detach the concept of sublimity from rhetorical devices and images and to redefine it as an aesthetic experience. According to Addison, sublimity was one of the "pleasures of the imagination," along with beauty and novelty.[8]

As an aesthetic principle, sublimity was unusual in that from the start it was associated less with works of art than with features of nature. Beginning with Addison, the sublime came to be considered a feeling aroused most notably by greatness and vastness in natural scenery—by mountains, deserts, precipices, oceans, and the like. Addison had taken the Grand Tour in 1699, and many of his examples of "the Sublime in external Nature" were based on memories of his Continental travels. During the eighteenth century the concept of sublimity developed along with a more general romantic enthusiasm for the wilder and untamed aspects of landscape—for (in the words of Horace Walpole) "precipices, mountains, torrents, wolves, rumblings. . . ."[9] A series of "excursion poets" are said to have "sent their imaginations on grand tours of the universe to marvel at all that was vast or grand—mighty continents with mountains and oceans, majestic rivers, subterraneous regions with caverns measureless to man."[10]

In the 1750s, Edmund Burke combined Addison's ideas and the literary tradition of the "natural Sublime" to develop his influential theory of sublimity.[11] In the two editions of *A Philosophical Enquiry into the Origin of our Ideas of the Sublime and Beautiful* (1757, 1759), Burke sharply distinguished the two ideas. As is usually the case when such bipolarities are proposed, Burke was arguing for the superiority of one of the two terms—in this case, for sublimity as the more intense and therefore more valuable experience. He disdained the traditional qualities of beauty, such as smallness, smoothness, delicacy, variation, and color. Such qualities, he suggested, may arouse aesthetic pleasure; however, this emotion is weak in comparison with that of sublimity. According to Burke's psychological analysis, the experience of the sublime rests on a delicate psychological equilibrium. Pain and danger "are simply terrible, but at certain distances, and with certain modifications, they may be, and they are delightful. . . . [Sublimity is] tranquillity tinged with terror."[12] To put it another way, sublime terror is aroused by anticipated, not actual, pain or danger. In its highest form, sublimity produces astonishment, "that state of the soul in which all its motions are suspended with some degree of horror."[13] The experience of sublimity therefore depends on the delicate equipoise of conflicting emotions. It is connected with pain and fear, but not too closely; it is defined by nervous tension, but not too much; it depends on danger, but only theoretical danger. Sublimity celebrates ambivalence.

Burke went on to describe the qualities of objects most likely to arouse such sensations: power, deprivation, vacuity, solitude, silence, great dimensions (particularly vastness in depth), infinity, magnificence, and finally obscurity (because mystery and uncertainty arouse awe and dread). In all cases the objects overwhelm the observer, making him feel small and powerless by comparison. Burke cited specific examples of natural scenes that arouse the idea of sublimity: the open ocean, the starry heavens, inarticulate cries of man or beast. Above all, darkness, the greatest of the "privations," is sublime; "Night increases our terror more perhaps than any thing else."[14] Thus, subterranean environments, being at once dark, deep, and deprived, are ideal examples of sublimity in nature. The obscurity and formlessness of the mine, which Mumford found so repugnant, are precisely the characteristics that make it, from Burke's point of view, so admirable.

Thus cave tourism began as a deliberate quest for sublime experience. In Europe and in the British Isles, eighteenth-century aesthetic pilgrims of the middle and upper classes descended into subterranean realms that had hitherto been the haunts of the enslaved and the wretched. In Italy these travelers particularly sought out the Blue Grotto of Capri and the Cave of the Sibyl near Naples. The sea cave of Capri was a gentle excursion by boat, but to retrace Aeneas's descent to Lake Avernus the tourists had to be carried on the backs of porters through what one of them described as a "black, repulsive pool"[15] of water. More attractive was the Cave of Adelsberg in Germany, well known for its transparent white pillars and brilliant stalactites. Like other such sites, it offered minerals and fossils at a souvenir shop. Continental travelers also visited the cave of Agtelek in Hungary and various fossil caves in Germany and Belgium, while British tourists took guided tours of Derbyshire caves (with musical accompaniment) or even descended into working mines. The most popular British cavern was Fingal's Cave on the island of Staffa, off the western coast of Scotland. Since its discovery in 1772, Fingal's Cave "had been invested with mythological overtones and . . . an almost religious significance" as an image of "the beginning of the world at the far ends of the earth."[16]

To be sure, mountains were the favorite natural source of sublime emotions.[17] British travelers began to consider the Alps as a part of, rather than as an obstacle to, the Continental Grand Tour. In the British Isles, the Peak District became a major resort area by the mid-eighteenth century, and by the end of the century the Lake District, along with more remote areas of Wales and Scotland, attracted many seekers of sublimity.[18] But mountains and caverns are complements, not opposites. What goes up also goes down; caves, hollows, precipices, cataracts, gorges, and crags—all gouges into the earth—are integral parts of most mountainous terrains, and were then regarded as among their most compelling features.

The most sublime mountains of all were volcanoes, which seemed conduits into the very bowels of the earth. All over Europe, from the 1750s onward, traveler-naturalists visited not only Vesuvius, which had long been a stop on the Grand Tour, but also the volcanic regions of Auvergne and Sicily. Interest in volcanoes had been aroused by a series of earthquakes in the 1750s and the 1760s (above all by the catastrophic Lisbon earthquake of 1755) and by the frequent, spectacular eruptions of Vesuvius during the 1760s and the 1770s. Sir William Hamilton, the British Envoy to the Court of Naples, climbed Vesuvius hundreds of times, taking measurements and writing descriptions which he sent as letters to the Royal Society and which were eventually published as a book. (He also collected some of the antiquities just then being unearthed from Pompeii.) In his reports Hamilton argued that "volcanoes should be considered in a creative rather than a destructive light."[19] They were not freaks of nature, but orderly and predictable phenomena that, by their successive eruptions, played an important part in molding the landscape. In this tone of admiration, Hamilton noted that the eruptions were most beautiful by night, saying of an August 1779 episode that "the fiery fountain, of so gigantick a size upon the dark ground abovemention'ed made the most glorious contrast imaginable, and the blaze of it reflecting strongly on the surface of the sea, which was at the time perfectly smooth, added greatly to this sublime view."[20]

As the case of Hamilton suggests, the discovery of sublimity in external nature played a key role in the emergence of modern geological concepts. The emotion of sublimity is, above all, related to perceptions of immense scale. The immensity of time, as well as that of space, arouses fear and awe, the shudder that arises from the apprehension of one's unimportance and impotence before grand natural powers. The appreciation of time's immensity developed along with the taste for images of natural sublimity. Vistas such as exposed strata or extinct volcanoes incarnated in stone the emerging geological principle of vast power over vast time. Because of the eighteenth-century upsurge of interest in topographical features, these images were noted more and more often by travelers. They also became more and more popular as the subjects of landscape paintings. As Marcia Pointon writes, "It was the coupling of Burke's theories of the sublime to landscape that really opened the way to a union between art and geology."[21] Although landscape painting was by no means new in the eighteenth century, it was a relatively lowly genre—particularly in comparison with history painting, which was supposedly concerned with more morally elevating subjects. By the nineteenth century, however, the search for sublimity had led to an alliance between landscape painting and geology. From that alliance emerged what John Ruskin called "historical landscape . . . landscape painting with a meaning and a use."[22]

The concept of sublimity is thus part of the cultural context of the discovery of geological deep time. It also is part of the context of the aesthetic discovery of industrial technology. The iconography of sublimity (the key image here is the exploding volcano) provides the link between the natural landscape and the technological one. For example, when Joseph Wright of Derby came to Naples to study painting in 1774, he witnessed an eruption of Vesuvius, which he made the subject of several powerful studies in sublimity. One of them, an oil painting, shows an imagined view of the interior of the crater, a chaos of rock towers, molten lava, explosions of fire, and surging gases. When Wright returned to England at the end of 1775 he continued to paint scenes of Vesuvius, but he also began to use volcanic images in painting the new industrial landscape. His perception of the sublimity of nocturnal fire is most successfully transferred to an industrial subject in his painting *Arkwright's Cotton Mills, by Night* (ca. 1782–83).[23]

We have already considered the oft-repeated association of industry with hell, beginning with mines and moving outward to factories and entire regions. In short, the vocabulary of sublimity was gradually but persistently transferred from nature to industry. This transference has ideological as well as aesthetic significance.[24] Sublimity, after all, was originally a rhetorical mode. The purpose of rhetoric is to devise effective ways of pleading, persuading, and arguing. Rhetoric is not a disinterested theory, but an "interested" technique that analyzes discourse as a form of power.[25] Burke in particular emphasized that feelings of the sublime are aroused by sights that make the observer feel small and insignificant. Sublime emotions are submissive ones. When the display of power is technological rather than natural, the sense of helplessness is aroused not by natural forces but by manufactured ones. The aesthetic of sublimity implies technological determinism.

Consider, for example, the series of mining disasters that occurred in nineteenth-century Britain: Felling (1812), Wallsend (1829), Haslam (1844), Risca (1862). In each case the explosions and fires were described by eyewitnesses as resembling those of a volcanic eruption. Here is a description of the Felling disaster, from a funeral sermon for the victims later published by John Hodgson, the clergyman who officiated:

The subterraneous fire broke forth with two heavy discharges. . . . A slight trembling, as from an earthquake, was felt for about a half a mile around the workings. . . . Immense quantities of dust and small coal accompanied these blasts, and rose high into the air, in the form of an inverted cone. . . . The dust, borne away by a strong west wind, fell in a continued shower from the pit to the distance of a mile and a half. . . . It caused a darkness like that of early twilight, and covered the roads so thickly, that the footsteps of passengers were strongly imprinted in it.[26]

Such images of subterranean eruption were by that time well established as examples of natural sublimity arousing a "pleasing horror" in distant observers. But for those who worked in the mine—who were not observers of the natural landscape, but participants in the industrial landscape—no such distancing was possible. Hodgson continues:

As soon as the explosion was heard, the wives and children of the workmen ran to the working-pit. Wildness and terror were pictured in every countenance. The crowd from all sides soon collected to the number of several hundred, some crying out for a husband, others for a parent or a son, and all deeply affected with an admixture of horror, anxiety, and grief. . . . By twelve o'clock, 32 persons, all that survived this dreadful calamity, were brought to day-light. The dead bodies of two boys . . . who were miserably scorched and shattered, were also brought up at this time . . .[27]

Humphrey Jennings, who quotes this description in his superb collection *Pandaemonium*, suggests that the stock comparisons of pit explosions to volcanic eruptions tend to make such disasters seem inevitable:

In the fantastic symphony of the Industrial Revolution from the beginnings up to today—yes, today—the dull subterranean explosions of the great and horrible pit disasters return (precisely like the periodic activities of a volcano) like a Fate theme, like reminders from the unconscious (as in dreams) of this work that goes on, out of sight, night and day. Yet these "accidents" are unnecessary, and the idea that they are due to "Fate" is a conception à la Calvin to depress the people.[28]

Whether or not the comparison of mine eruptions to volcanic ones was intended "to depress the people," as Jennings suggests, it would have that effect. This is the same sense of fatalism expressed by Dickens when he compares the laying of the railroad track through Camden Town to an earthquake. In both cases the comparison implies that progress cannot be stopped. Mechanical force is both insuperable and inevitably destructive. The observer submits to the spectacle of sublimity with awe and astonishment. The spectacle is active, the observer passive. Politics depends on speech, but the experience of the sublime renders one speechless with horror and amazement. Sublime technology is autonomous technology, technology-out-of-control.[29]

The hellish image of erupting flames links subterranean and technological sublimity. It is not the only connection, however. In his *Enquiry*, Burke suggests that feelings of sublimity might be aroused not only by external nature but also by the "artificial infinite," a nonorganic visual object of great dimensions. According to Burke, such objects give the sensation of sublimity because of the physiological response they demand. When the eye moves from one small object to another, it experiences moments of relaxation. When an object is both simple and vast, however, the eye cannot rest. The image is always the same, and seems to have no bounds. The eye is therefore cast into a state of tension, and the mind experiences the sensation of sublimity.[30] A long colonnade, or any large plain object, might by its uniformity and size arouse the impression of the "artificial infinite."

In the first industrial revolution, the image of nocturnal eruption is the pivotal point where subterranean imagery was transferred to technology in general. In the second industrial revolution, that point of transfer was more often the image of a vast, illuminated underground space, stretching off into infinity. This second image is potentially far more enticing than the older, diabolical one. During the nineteenth century, the sublimity of artificial infinitude gradually displaced that of hell as the prevailing aesthetic of the underworld.

The ideal of constructing an awe-inspiring space of great dimensions long precedes the nineteenth century and even Burke. This ideal in fact shaped Western architecture beginning with the Roman Empire. According to the architectural historian Sigfried Giedion, the supreme goal of Roman builders was to hollow out a large and seemingly indestructible interior space. The construction techniques developed by the Romans—the arch, the vault, and structural concrete—allowed them to build, for the first time, interior space on a monumental scale. The dome of the Pantheon (begun in A.D. 118–119), with its "exclusive concentration upon interior space,"[31] represents a breakthrough to a new conception of space—one that was to dominate architecture up to the beginning of the twentieth century. "From then on," writes Giedion, "all concepts of architectural space would almost invariably be synonymous with the concept of a hollowed-out interior."[32] Giedion traces the development of interior space through a series of prototypical buildings: San Vitale in Ravenna, S. Maria del Fiore in Florence, St. Peter's in Rome, St. Paul's in London, the Pantheon in Paris.[33]

Once again, an aesthetic principle carries ideological significance. The architectural ideal of a centralized, enclosed structure expresses an urban ideal. The enclosed space represents the authority of civilization, its independence from nature, and its permanence and solidity in the face of nature's flux. "The Pantheon," Giedion notes, "is a completely self-contained structure; its circular form represents a closed system that demands isolation."[34] Interior and exterior—civilization and nature—are idealized as separate systems. The purpose of the interior space is to instill a sense of awe before this triumph of the social world: the sublime emotion of submission in the face of vastly superior power.

Nature, however, could not be totally banished from these monumental interiors. In the days before artificial illumination, the interior space that was completely sealed off would also be completely dark. Giedion notes that in the Pantheon "only the sky is granted free access." The open oculus at the very top of the dome, the sole link between the interior and the external world, admits a constantly shifting beam of light as the sun passes over the building.

In subsequent centuries, in other buildings, interior walls were increasingly perforated to admit more light.[35] When more light entered, however, the awe-inspiring sense of enclosure was diminished. The architectural ideal of a vast, enclosed, yet luminous interior space was a natural impossibility. It could be realized only through the power of imagination or through the power of technology.

Huxley said that half-waking dreams presaged nineteenth-century science; they also presaged nineteenth-century technology. The aesthetic possibilities of artificial light—light so extensive and brilliant that it could illuminate a large underground space—were explored in imagination long before the electrical technologies enabled them to be realized in fact. One of the most influential of these aesthetic explorers was Giovanni Battista Piranesi (1720–1778). Piranesi, of course, drew actual Roman buildings with painstaking realism in the collections familiarly known as *Views of Rome* or *Antiquities of Rome*.[36] In his "imaginary prisons"[37] engravings (published in 1745 and issued in a second edition in 1761), he turned from actual constructions to imaginary ones. The same architectural themes persist—especially Piranesi's obsession with enclosed space and with "dreadful solidity,"[38] which echoes the obsession of the Roman builders themselves. The imaginary prison buildings, however, are at once sealed and illuminated. In Marguerite Yourcenar's vivid phrase, the prisons seem "struck by the rays of a black sun." In these surreal spaces, she writes, "time [does not] move any more than air; the perpetual *chiaroscuro* excludes the very notion of the hour."[39]

The imaginary prisons of Piranesi are not literally underground, of course, but they convey an overwhelming sense of enclosure. They are architectural approximations of a subterranean environment. More than any realistic building, they embody the ideal of artificial infinity. In Yourcenar's words: "Animal and plant are eliminated from these interiors where only human logic or human madness rules; no trace of moss touches these bare walls. The natural elements are absent or narrowly subjugated: earth appears nowhere, covered over by tiles or indestructible pavings, air does not circulate. . . . a perfect immobility reigns in these great closed spaces. . . ."[40] In the corners of Piranesi's prisons crouch various tools and machines: gibbets, ladders, trestles, jacks, scaffolds. They could be ordinary construction devices, but in the context of these

One of Piranesi's "imaginary prisons" engravings. From Piranesi, *Oeuvres choisies* (Paris, 1913). Courtesy of Rotch Library, Massachusetts Institute of Technology.

gloomy spaces they irresistibly suggest torture and execution. Here are the sublime references to pain and danger, but without darting flames, smoky eruptions, or clangs of metal. This is a spooky, motionless, silent realm from which nature has been banished.

Piranesi claimed that the prison images came to him in 1742 during an episode of delirium resulting from a malarial fever (a common affliction in the campagna, where he worked). By definition irrational, the image of an illuminated underworld continued to be associated with a dreamlike state of mind. Samuel Taylor Coleridge, in his famous preface to "Kubla Khan" (subtitled "A Vision in a Dream. A Fragment"), explains how the writing of the poem came about: Having taken "an anodyne" to relieve "a slight indisposition," Coleridge fell asleep while reading about the Khan Kubla's palace and garden. In his dream he imagined that the palace had sunk underground, and that the walled-in garden had been transformed into a dream landscape of ice, water, and light:

Where Alph, the sacred river, ran
Through caverns measureless to man
Down to a sunless sea. . . .

It was a miracle of rare device,
A sunny pleasure dome with caves of ice!

Coleridge had seen Piranesi's "imaginary prisons" engravings, and he recalled their supposed origin in a feverish delirium.[41]

As Coleridge showed, the fantasy of an illuminated underworld could be expressed in poetic lines as well as in engraved ones. Edgar Allan Poe's poems "Dream-Land" and "Ulalume" appear to be set in Belzubia, an inner world illuminated by a mysterious, diffuse light from the highly refracted rays of the sun and moon entering through the polar opening.[42] In "Rêve parisien" [Parisian Dream], part of the collection *Fleurs du mal,* Poe's admirer Charles Baudelaire describes his dream of a "terrible landscape" from which he, the poet, had banished "irregular vegetation" in favor of an "intoxicating monotony/Of metal, marble and water." In the "endless palace" of Baudelaire's imagination, "sleeping pools" were surrounded not by trees but by colonnades, and heavy waterfalls hung like crystal curtains from high walls of metal:

Architect of my own fantasies,
I made pass, at will,
Under a tunnel of precious stones
A conquered ocean; . . .

There was no star, no vestige,
Of a sun, even at the horizon of the sky,
To illumine these prodigies
Which shone with a personal fire![43]

We started by considering the ancient tradition of the underworld as ugly, repulsive, slimy, dark. The reconceptualization of the underworld as having either the diabolic sublimity of fire or the mysterious sublimity of "artificial infinity" represents an intermediate stage in which revulsion and attraction were mingled. Now we are about to enter yet another stage in this aesthetic transvaluation—one in which revulsion was left behind and the underworld was perceived as a magical paradise. By the middle of the nineteenth century, the highly ambivalent emotions aroused by subterranean sublimity had begun to yield to unambiguous appreciation of subterranean beauty. The image of artificial infinity gradually shed its aura of terror and assumed the mantle of enchantment.

Burke had declared that sublimity and beauty were incompatible experiences: the first was a highly ambivalent state of nervous tension, the second an unambiguous state of pleasure in which the nerves are relaxed. But in "Rêve parisien," Baudelaire blended these experiences. The landscape is sublime in its "terrifying novelty," but above all it is beautiful. To be sure, this beauty does not comport with the supposedly objective standards inherited from Greek philosophers and ridiculed by Burke—standards such as proportion, fitness, harmony, smallness, smoothness, delicacy, variation, and color.[44] The essence of this new, inorganic beauty is the replacement of the light of day by the gleams, glows, and reflections of artificial light. The sounds of life are replaced by (in Baudelaire's words) "a silence of eternity." The only organic element left is water, which assumes the quality of liquid metal.

This is beauty that depends upon the banishment of nature—and that means the banishment of man. Mumford used the mine as the image of paleotechnic industry, and a mine cannot be imagined without miners. Most caves, however, are uninhabited. Seekers after the sublime sought out empty caverns far more often than working mines. In the fantasy underworld the only human presence is the omnipotent observer, who has the place all to himself. The human presence would be an unwelcome reminder of the organic in a realm of inorganic beauty.

A telling example of the effect of human presence in the underworld is found in an 1827 letter from the English actress Fanny Kemble to a friend. The letter describes a recent visit by Kemble, her father, and some friends to the Thames Tunnel, then being dug by the Brunels. The fact they made this descent is itself significant. Their visit was a technological version of the cave tours that had become popular during the eighteenth century; it was also an early example of the middle-class descent into the urban underworld that became a common experience later in the nineteenth century, especially in the form of subway rides. At first Fanny Kemble was surprised to find the subsurface environment pleasing rather than fearsome. At the bottom of the stairs leading down into the tunnel works, she turned around and admired what Burke would call an artificial infinity:

. . . as far as sight could reach stretched a vaulted passage, smooth earth underfoot, the white arches of the roof beyond one another lengthening on and on in prolonged vista, the whole lighted by a line of gas lamps, and as bright, almost, as if it were broad day. It was more like one of the long avenues of light that lead to the abodes of the genii in fairy tales, than anything I had ever beheld. The profound stillness of the place . . . to which the vaulted roof gave extraordinary and startling volume of tone, the indescribable feeling of subterranean vastness, and amazement and delight I experienced, quite overcame me. . . .[45]

This is not a demonic underworld; it is a vast, illuminated, silent fairyland. However, the images and Kemble's response begin to change. She tells how Brunel asked her father if the party wanted to penetrate further to where the workers were employed:

. . . an unusual favour, which of course delighted us all. So we left our broad, smooth path of light, and got into dark passages, where we stumbled among coils of ropes and heaps of pipes and piles of planks, and where ground springs were welling up and flowing about in every direction, all which was very strange. As you may have heard, the tunnel caved in once, and let the Thames in through the roof. . . .

Having left the "path of light" for a dark, strange underworld, the party enters the realm of labor:

. . . the appearance of the workmen themselves, all begrimed, with their brawny arms and legs bare, some standing in black water up to their knees, others laboriously shovelling the black earth in their cages (while they sturdily sung at their task), with the red, murky light of links and lanterns flashing and flickering about them, made up the most striking picture you can conceive. As we returned I remained at the bottom of the stairs last of all, to look back at the beautiful road to Hades. . . .

Fanny Kemble closes the letter with this wonderful remark: "I think it is better for me, however, to look at the trees, and the sun, moon, and stars, than at tunnels and docks; they make me too humanity proud."

At one end of the tunnel, the beautiful road; at the other end, Hades—contrasting images of the underworld that are also contrasting images of technology. The first image of subterranean sublimity, that of hell, is compellingly suggested by paleotechnic industry, by the steam and fires and grime and noise of the first industrial revolution. Here, Fanny Kemble advances one step further (both literally and figuratively) and makes a connection between the landscape and the workers. She looks not only at the subterranean environment but also at the people who work there, and she begins to imagine what it might be like to labor in the muck. If, like most middle-class visitors, she had been confined to the public parts of the tunnel—the "beautiful road"—she would have seen no reminders of labor or floods. In that case, she would have felt as if she were visiting a natural cave rather than a tunnel excavated by human labor. She could have interpreted the tunnel as a fantastic fairyland, like "the abodes of the genii in fairy tales." Only when she penetrated into the working part of the excavation was she forced to remember the labor that had dug it.

Diabolical images of sublimity are inseparable from labor, from the sight of the shadowy figures silhouetted by the flames of production. The aesthetic pleasures of technological sublimity had always been tainted, so to speak, by the human presence. In the sublime images of artificial infinity, however, labor can be banished. The aesthetic fantasy is closely related to the social fantasy of eliminating class conflict, of exploiting nature without exploiting people. History is banished along with nature. Let us see how these two fantasies are connected in some later imaginary underworlds.

Two such novels appeared nearly simultaneously on opposite sides of the English Channel: Edward George Bulwer-Lytton's *The Coming Race* (published anonymously in 1871) and Jules Verne's *Vingt mille lieues sous les mers* [Twenty Thousand Leagues under the Sea] (composed 1865–1869, published in serial form 1869–1870).

Bulwer-Lytton's highly popular novel *The Last Days of Pompeii* had been published several years earlier, in 1866. *Pompeii* had told the story of a buried past (it concluded with a description of recent excavations there).[46] *The Coming Race* imagines a buried future. The narrator of *The Coming Race* becomes acquainted with a mining engineer who has detected strange lights and noises in a deep chasm. When the two of them return to investigate, the engineer is killed in a fall. To the narrator's horror, the body is snatched by a gigantic reptilian monster. As he follows the creature downward, however, the narrator finds himself in an underground realm more utopian than reptilian. The chasm leads to a lamplit road amid a Baudelairean landscape, with fields covered with a strange vegetation (golden red, not green), and with lakes and streams curved into artificial banks ("some of pure water, others that shone like pools of naphtha") under a dome sunless but bright. The narrator comes to a vast building, partly hollowed from a great rock, which is fronted with columns and ornamented with fantastic capitals reminiscent of early Egyptian architecture. There he finally meets an inhabitant of this underground realm, one of the vril-ya—a race of tall, winged creatures.[47] Terrified, the narrator faints. When he awakens, he finds himself in surroundings that perfectly blend comfort, gadgetry, and beauty. He is in "an immense hall, lighted by the same kind of lustre as in the scene without." The air in the room is filled with sweet smells and soft music. An automaton stands in the corner, ready to run errands at the touch of a vril-ya's staff. After the narrator has been served a delicious meal, he is led onto a platform, which lifts him to a high balcony. There he gazes upon a landscape "of a wild and solemn beauty impossible to describe,—the vast ranges of precipitous rock which formed the distant background, the intermediate valleys of mystic many-coloured herbage, the flash of waters, many of them like streams of roseate flame, the serene lustre diffused over all by myriads of lamps. . . . so splendid was it, yet sombre; so lovely, yet so awful."[48]

The vril-ya are as sublime as their landscape. They are at once handsome and frightening in their "calm, intellectual, mysterious beauty."[49] The source of their power is technology. The name of the race comes from its command of vril, a force that resembles electricity "except that it comprehends in its manifold branches other forces of nature, to which, in our scientific nomenclature, differing names are assigned, such as magnetism, galvanism, &c."[50] (The narrator quotes Faraday as an authority for the belief that these forces are interdependent and convertible.) With this all-purpose technology, the vril-ya light lamps, cut through rocks, fly, heal diseases, and slay reptiles such as the one that molested the body of the hapless engineer. Liberated by vril from competition and want, the vril-ya live in a paradise of technological wizardry and artistic taste.

Captain Nemo's submarine in *Twenty Thousand Leagues under the Sea* is another such sunken paradise. Admittedly it is cheating a bit to call the *Nautilus* an underworld fantasy, when in fact it is underwater. If we wanted to be strict about these distinctions, we would have to confine ourselves to Verne's descriptions in *Journey to the Center of the Earth,* where the travelers find, in addition to some prehistoric creatures, an immense cavern glowing with a mysterious light:

. . . the illuminating power of this light, its quivering diffusion, its clear and dry whiteness, its low temperature, its lustre in reality superior to that of the moon, evidently suggests an electrical origin. It was like an aurora borealis, a continuous cosmic phenomenon that filled this cavern capable of holding an ocean.[51]

Or we could dwell upon the beauty of the cavern, also immense and illuminated, beneath *L'île mystérieuse* (1875), where the *Nautilus* finds its final resting place. There the lofty vaulted roof, the irregular arches, and the side cliffs are all "flooded with the electric fluid, so that the brilliancy belonged to them, and as if the light issued from them."[52] Or we could cite the man-made underworld of *Black Indies,* the utopian Coal Town, where "numerous electric disks replaced the solar disk" and flooded the cavern with intense light:

When this underground town was lighted up by the bright rays thrown from the discs, hung from the pillars and arches, its aspect was so strange, so fantastic, that it justified the praise of the guide-books, and visitors flocked to see it.[53]

Still, Verne always considered *Twenty Thousand Leagues under the Sea* one of his most significant works, and in that sunken world we find a classic expression of the aesthetic of technological magic that emerged along with the second industrial revolution. When Professor Aronnax and his two companions awaken below, they are (as is typical in these stories) taken on a tour of their new environment. Doors open magically before them (this detail is also typical) as they are taken to the dining room, where they find a French bourgeois paradise: heavy oak furniture, china and crystal and silver, exquisite paintings, gourmet food.[54] After dinner, other doors open to lead the three visitors to a library that, in Aronnax's words, "would do honour to more than one of the continental palaces," and finally to a drawing room in which "an intelligent and prodigal hand had gathered all the treasures of nature and art."[55]

Verne's account of the *Nautilus*'s artistic and natural wonders is immediately followed by a description of its technological wonders in the famous chapter XI, "All by Electricity." Here Captain Nemo tells how electricity provides communication between the *Nautilus* and its tender, fuels the stove, distills and heats water, and powers the vessel. Some of the technical details are implausible, "but," as one critic has commented, "none of that seems important. It's the atmosphere he creates that counts."[56] In this atmosphere, what had seemed magical now seems technically plausible. Furthermore, what was technically possible now seemed magically beautiful. In the hellish mills of the first industrial revolution, technology and art were usually assumed to be opposites. In the subsurface paradises of the later nineteenth century, they coexist in perfect harmony.

And so do people coexist in perfect harmony. That is because they are blessed by an apolitical and classless source of technological power that wonderfully extirpates social conflict. If vril resembles electricity, it can just as well be said (at least for Verne's earlier novels) that electricity resembles vril. They are both technologies that permit direct dominion over nature without the mediation of human labor. The aesthetic fantasy of an underground paradise is congruent with a particular social fantasy: the possibility of social transformation through a crucial discovery. In Bulwer-Lytton's case the fantasy is "a projection of the idealized social attitudes of an aristocracy."[57] In Verne's case it expresses a middle-class, technocratic dream of machines that exploit nature without exploiting men.[58] Both writers describe an artificial environment where technological and artistic beauty coincide, and where social conflicts have been resolved thanks to the definitive conquest of nature. The aesthetic fantasy of an artificial infinite is tied to a particular social fantasy: that of a community without class struggle or exploitation, a social vista as monotonous and vast as an endless colonnade.

Let us confirm this analysis by looking somewhat more briefly at two more underworld paradises, again from opposite sides of the Channel but this time from the 1880s.

William Delisle Hay's novel *Three Hundred Years Hence* (1881) develops the themes described above in a particularly unpleasant way, for its social fantasy is violently racist. The book is supposedly written in 2180 A.D. by a professor holding the chair of English Antiquities and Literature at the Historical College in the City of Londinova, in the State of Atlantis. Back in the late nineteenth century, the good professor tells us, the world's population began to increase exponentially; it now totals 130 billion. In order to cope with this demographic nightmare, humanity has had to undertake stupendous engineering projects: building submarine and subterranean cities to provide living space, farming in caverns, and the like.

The professor then describes an underground journey to the Interior, where 21 billion people live in ten underground states. He descends 50 miles below the earth's crust in a "terra-car," a sort of luxurious subway that travels through tunnels made by an earth-boring machine using the power of "basilico-magnetism." Like vril, basilico-magnetism is a kind of all-purpose energy source based on vague analogies with magnetism and electricity. The narrator is happily surprised when he arrives at the subterranean city of Argenta: "Instead of the murky gloom you might expect, but feebly lit by a lamp here or there, you have a broad clear atmosphere, quite as bright as you left in the upper world. Above is a sort of luminous haze that shuts out any glimpse of the rocky roof miles overhead. . . . Numberless twinkling points all round indicate the sources of light, but you are not conscious that the light proceeds entirely from them."[59] Down below is a world's fair of architectural styles: Chinese pagodas, Muslim mosques, Grecian temples, and Swiss chalets all hang from the cliffs and rocky platforms.[60] This interior realm shows "no trace of vegetation, no sign of that softer loveliness of the sunshine," but its beauty surpasses that of nature. In this "rich metropolis of the metals," the profusion of "towers and domes, minarets, cupolas, pyramids, and obelisks" is "so utterly grand, so beautiful in their glory of art, that one feels never tired with the pleasures of the sight."[61]

In time the professor reveals that social conflict has been eliminated through an Epoch of Final Wars, ending in a Great Social Revolution that ushered in a Century of Peace. The universal fraternity that reigned during the Century of Peace gradually gave way, however, to a more realistic appreciation that the yellow and black races were incapable of civilization. "Tender humanity shuddered as it perceived the inevitable conclusion,"[62] but the feelings of guilt were removed when the Mongols seceded from the Union of Humanity and when the Negroes in South Africa rose in insurrection against it. In both cases "the consummation of the unwelcome but unavoidable task was at once proceeded with,"[63] and the yellow and black races ceased to be. These final solutions to class and racial conflict opened the way to the splendid achievements of three hundred years hence. Like Bulwer-Lytton and Verne, Hay assumes that technological and artistic achievements are inseparable. Like them, too, he assumes that they are made possible when the historical world of inequity and strife is left behind.

It is a relief to turn from Hay's genocidal fantasy to the far more benign vision of Gabriel Tarde's *Underground Man,* written in 1884 but not published until 1896. Tarde's story begins much like Hay's. A century and a half of vicious warfare has led to a universal peace in which invention flowers. The global state that now rules the world initiates a grand public-works project to rebuild Constantinople on the site of ancient Babylon, with vast irrigation works and with Abyssinian waterfalls harnessed to produce electricity. But Tarde, unlike Hay, is wholly ironic in contemplating such projects. "This imperial orgy in bricks and mortar,"[64] he tells us, bankrupts the state many times over. Even worse, technology proves helpless when a real calamity strikes. The sun begins to cool off. As a chill settles over the earth, long-dormant glaciers revive and move south. "A moving cliff composed of rocks and overturned engines, of the wreckage of bridges, stations, hotels and public edifices"[65] marches inexorably onward, as people flee before it to seek warmer climates. The last few survivors of humanity, facing extinction, huddle near the site of ancient Babylon. They decide that the only way to save themselves is to take refuge inside the earth. Like Captain Nemo, the "neo-Troglodytes" methodically gather the cream of human culture from museums, libraries, and industrial exhibitions. They dig out subterranean galleries, which they line with these master-pieces ("the true capital of humanity"); then they themselves enter the galleries, sealing themselves off forever from the surface of the earth.

Like Bulwer-Lytton's and Hay's narrators, Tarde's neo-Trog-lodytes are amazed to find themselves in an underground paradise: "They expected a tomb; they opened their eyes in the most brilliant and interminable galleries of art they could possibly see, in *salons* more beautiful than those of Versailles, in enchanted palaces, in which all extremes of climate, rain, and wind, cold and torrid heat were unknown; where innumerable lamps, veritable suns in bril-liancy and moons in softness, shed unceasingly through the blue depths their daylight that knew no night."[66] In the temperate sub-terranean climate they need little clothing, although they do enjoy iridescent coats of asbestos spangled with mica. For shelter, anyone can drill a hole in a rock. Architects, now called excavators, dig a wanton and free series of burrows, an "artificial and truly artistic landscape." The underground people eat meat frozen in the ice and obtain heat from cauldron-volcanoes. Outside their cities of art there is no countryside, only wilderness. They can skate or bicycle through fantastic criss-crossing galleries of crystal carved through seawater turned to ice and lighted by millions of electric lamps mirrored in emerald-green icicles. Freed from organic life and historical change, art and society blossom at last.

If a "remnant of the mythological" lurks in the corridors of nine-teenth-century science, more than a remnant lurks in the corridors of nineteenth-century technology, which are flooded with magical light. Electricity realized the ancient ideal of an artificial infinity, a hushed, awe-inspiring refuge. More than that, electricity seemed to realize the related fantasy of a labor-free source of power. It seemed to offer refuge not only from nature's strife but also from social strife. This is a utopian vision of the technological environment. Was this just a vision, a drug-induced dream, or did it correspond to anything in the waking life of the nineteenth century? And has that vision been borne out in any way in our own century?

Let us begin with the first question and compare these technological utopias with nineteenth-century realities. In particular, we need to compare subterranean illumination in fantasy and in fact. A key technological emphasis in the literary fantasies is the possibility of illuminating vast underworld enclosures. Such large-scale illumination was by no means common when Bulwer-Lytton, Hay, and even Verne and Tarde were writing. Indeed, there is a painful disparity between their brilliant fantasies and the blackness of actual underground environments.

Better lighting had long been needed for subsurface construction projects. Nineteenth-century tunneling works were lighted by candles or, somewhat later, by gas and oil lamps. These devices were a prime source of the discomforts and hazards of tunneling. They contributed to the sometimes extreme heat inside deep tunnels (temperatures reached 130°F during the construction of the St. Gotthard Tunnel, for example), and in those poorly ventilated areas they made the air sickeningly foul. Later in the nineteenth century other alternatives were tried. Inside the Brooklyn Bridge caissons, for example, sperm candles on iron rods were planted alongside the walkways. When the air became intolerable, the wicks and candles were made smaller, alum was mixed with the tallow, and the wicks were soaked in vinegar. Roebling also experimented with calcium lamps, or limelights, which burned a combination of compressed oxygen and coal gas (ordinary street gas would have been much less expensive, but it overheated the caisson).[67]

Electrical alternatives had been considered ever since the principle of electric arc lighting had been discovered early in the century. Only in the 1870s, however, when better generating machines became available (notably the Gramme dynamo), did arc lighting become a reasonably efficient source of illumination for large industrial installations.[68] One of the first major subterranean projects to use both arc lamps and filament lamps was the Severn Tunnel, dug between 1873 and 1886.[69] Still, for some projects the problems of providing electrical power underground seemed greater than the potential benefits. For example, when the caissons for the Forth Bridge were sunk in the 1880s, "Lucigan lamps"—brilliant, naked flames over 3 feet long, produced when creosote oil was forced through a small nozzle by means of air pressure[70]—were used instead of electric lamps.

Electricity came slowly to the coal mines, where the need for stronger, safer lighting was most pressing. The illumination there was so poor that some miners developed nystagmus, an eye disease caused by prolonged strain and use of peripheral vision in the near dark. The usual sources of illumination—candles and lamps—were likely to set off explosions of firedamp or coal dust. (Firedamp is a mine gas that is explosive when mixed with air in proportions of between five and fifteen parts per hundred.) In coal mines, naked flames, sparking, and even transitory incandescence must be avoided or carefully controlled. When extensive mining began in the deep, gassy pits of Northumberland, Durham, and Cumberland in the early nineteenth century, many explosions resulted from the use of candles there.

The first reasonably safe source of illumination in coal mines was the Davy lamp, devised in 1815 by Sir Humphry Davy, who was motivated by the terrible explosion at Felling Colliery. He found that a cylinder of wire gauze placed entirely around the flame of an oil lamp made it safe in most conditions of firedamp; the gauze distributed the heat of the flame over its surface area so no one point of the gauze would get hot enough to cause an explosion. During the course of the nineteenth century the Davy lamp was improved both in safety and in illuminative power, but its light remained feeble (about half a candlepower).[71]

Electric filament lamps were first tried in the coal mines of Lanarkshire, Scotland, in 1881. Joseph Swan designed a fixture in which the bulb was immersed in water, so accidental breakage would extinguish the filament at once. Soon he replaced this arrangement with a protective outer bulb of gas. In 1883 Swan devised battery-powered electrical handlamps—forerunners of the flashlight. Safety lamps, based on Davy's principle, are still used for gas testing, for unlike the handlamps they indicate the presence of firedamp. Electric headlamps provide much brighter illumination for the miners, however, and they have become standard equipment.

Electrical lighting of the coal face was not permitted in Britain until 1934, when fixed electric lights began to be installed on the roadways. Caution was necessary because the presence of any electrical apparatus in coal mines requires stringent precautions to prevent flames and sparks. The worst coal-mine disaster in British history occurred at the Universal Pit in Mid Glamorgan (Wales) in 1913, when a firedamp explosion was ignited from sparking caused either by falling rocks or by electrical signaling apparatus.[72]

It is not in the working environment of the underworld, which remains dangerous and dark to this day, but in the consumer environment of tourism that we find the earliest approximations of the supernaturally illuminated underworld. Once again the uninhabited cave, not the working mine, allowed fantasy to flourish most freely. Innovative and spectacular subterranean lighting was first introduced (and is still mainly found today) in commercial caverns. Although few American caves were opened to general tourism before the 1920s, when automobile travel made them far more accessible than before, the electrification of caves had begun much earlier. The Luray Caverns, in Virginia, were electrified in the 1880s, when only a half-dozen or so caves were open to the American public.[73]

Even before most of the tourist caverns were electrified, however, promoters on both sides of the Atlantic had experimented with innovative lighting techniques. As early as 1849, when members of the British Association for the Advancement of Science visited the limestone caverns of Dudley, they were treated first to a lecture on the formations and then to this spectacle: ". . . red and blue fires were lighted at various parts of the caverns, the effect of which was striking and magnificent in the extreme, and drew forth shouts of admiration from the crowds who thronged the caves; and, as each successive blaze revealed the extent and form of the place, lighting up the projections and angles of the rocks, scenes of indescribable grandeur were produced."[74] (The report adds, though, that the visitors sought fresh air because of the "sulphureous vapours arising from the burning of the coloured fires.") In the 1870s, tourists to Kentucky's Mammoth Cave—and these included many visitors from abroad—were treated to a "drama of night and day" acted out with their hand-held oil lanterns. In one chamber, where light-colored stones on the ceiling gave the impression of a starry sky, the guide slowly gathered up the lanterns and withdrew with them, as if a storm was covering the sky. Then he disappeared entirely, leaving the tourists to think truly sublime thoughts in the utter darkness. (One pondered, "Such must have been the primal chaos before space was, or Form was, or 'Let there be light!' had been spoken.") When the guide returned, carrying the lanterns, he imitated the cry of a rooster, causing the same visitor to proclaim it "one of the finest artificial sunrises that could possibly be produced." Later in the tour, in order to light up a large chasm called the Maelstrom, the guide tossed into it "Bengal lights"—a mixture of potassium nitrate, sulfur, and realgar that gave a bluish light and was more commonly used in signaling and in theater illumination.[75]

In brief, the consumers' environment of the cave, not the producers' environment of the mine, comes much closer to realizing the fantasy of an enchanted, illuminated underworld. But we miss the point if we search for congruence only in consumer environments that are actually underground. More often the literary fantasy has been realized in environments that are literally above the ground but that, in their splendid isolation from natural and social disorders, mimic underground conditions.

The crucial distinction is not between surface and subsurface environments, but between producer-based and consumer-based utopias. In an important 1970 article, "The Mythos of the Electronic Revolution,"[76] James W. Carey and John J. Quirk have shown how the development of electrical networks gave rise to a utopian vision of regional decentralization. This is primarily (though not exclusively) a vision of production. They cite a whole series of social prophets—among them Patrick Geddes, Peter Kropotkin, Gifford Pinchot, and Lewis Mumford—who forecast that electrification would simultaneously promote prosperity, community, and harmony with nature. According to Carey and Quirk, the vision of a technological utopia began in eighteenth-century America when "the rhetoric of the technological sublime forecast that mechanization and industrialization would not produce the untoward consequences apparent in the European version of the Industrial Revolution."[77] In the latter part of the nineteenth century, however, when it became clear that "America was not to be exempt from European history," "many intellectuals merely transferred notions of technological bliss from mechanics to electronics, and from the 1870s to the 1970s there has flourished a genre we have termed the 'rhetoric of the electrical sublime.'"[78]

The evidence already presented here suggests several modifications of this thesis. First, the advent of practical electrical technologies inspired Europeans as well as Americans with "notions of technological bliss." More important, the older rhetoric of mechanics was not simply transferred to electronics. The rhetoric itself changed significantly. The confusion arises from Carey and Quirk's indiscriminate use of the term *sublime*. One set of sublime images—those featuring the nocturnal fire of hell and volcanoes—was early transferred from nature to technology and became a significant vocabulary in interpreting the images of the first industrial revolution. A second set of images—those of "artificial infinity"—slowly moved from architecture to art to literature, from realistic constructions to fantastic illuminated enclosures. By the 1870s, the frightening aspects of this second type of sublimity had largely faded. Pleasing images of technological magic displaced (although they never entirely replaced) the more frightening ones of diabolical sublimity. The technological-aesthetic fantasies of Bulwer-Lytton, Verne, Hay, and Tarde do retain some sublime aspects, but they are primarily soothing, splendid underworlds.

But the main point of Carey and Quirk's article—and certainly a more significant point—is to contrast rhetoric with reality, "notions of technological bliss" with historical actuality. They cite the many predictions that electricity would simultaneously decentralize living patterns, production, and social power, so that grimy industrial cities and huge factories would give way to small, pleasant villages where farmer-craftsmen would carry out small-scale but highly efficient production. Carey and Quirk, along with Thomas Parke Hughes in a recent article,[79] stress the disparity between these apparently glowing predictions (an appropriate adjective here) and historical actuality. They agree that the predictions proved wrong because of a mistaken faith in technological determinism—or, more precisely, a mistaken faith in a deterministic ideology of technological progress. Instead, they concur, the development of electrical technology was greatly influenced, if not determined, by social and historical factors favoring continued centralization.

Hughes asks "Why did the visionaries of the second industrial revolution predict so poorly?" He responds to his own query that "persons and institutions with stakes in the cities, accustomed to organizing and controlling large-scale production in factories, adapted electricity to their ends."[80] With these people in charge, Hughes continues, electricity was used to promote urban concentration rather than regional decentralization: "Electricity made possible elevators and interior lighting, thereby concentrating population in high-rise buildings. Streetlights illuminated dark city canyons. Electrically driven mainline railroads and subways brought additional hordes of workers through tunnels into the cities. Dank subways stood as cruel mockery of the sunlight and fresh air electrification had been supposed to make available."

Hughes's description of modern city life—"high-rise buildings . . . dark city canyons . . . tunnels . . . dank subways"—evokes the ugly, dark underworld of ancient tradition. He contrasts the regionalist vision of sunlight and fresh air with the dark, centralized realities of modern urban life. But one could describe modern metropolitan life in much more favorable terms. One could dwell on the brilliance of department stores, the elegance of the "corporate spaceships" called office buildings,[81] the bustle of glamorous streets, the speed and convenience of subways that are not always dank but can be (like the RER lines in Paris) glittering high-tech showplaces lined with shops and restaurants. These examples too show that "persons and institutions with stakes in the cities" knew how to "adapt electricity to their ends"—commercial ends.

History has indeed failed to realize the regionalist utopia of small-scale, scattered, but efficient producers. The vision of regional decentralization, however, is only one part of "the mythos of the electronic revolution." The subterranean fantasies we have examined here embody another myth: that of an enclosed realm of consumer bliss, sealed off from the disorders of nature and society. If historical fact mocks the former dream, it has borne out the latter one.

Baudelaire is the key figure here. More than anyone else in the middle of the nineteenth century, he perceived that real cities, as centers of consumption, were assuming an aura of fantasy. The "Rêve parisien," the mysteriously lighted, silent, inorganic enclosed city of metal and water, is a *dream* city, but it is a *Parisian* dream. Baudelaire's Paris is a surreal, allegorical city, a hermetic temple of magical correspondances. It is not literally underground, but in its dreamlike self-enclosure and separation from nature it is, in the words of Walter Benjamin, "a sunken city" (though, Benjamin adds, "more submarine than subterranean").[82] The idea of a mysterious, hidden Paris, rich with hidden meanings and connections, was by no means new with Baudelaire. Even in the first part of the nineteenth century, the city was being invested with mythological stature.[83] Baudelaire, however, was obsessed with the city's mythic significance, and that obsession was closely linked to the technological, political, and social restructuring of Paris in his day.

That linkage is the subject of Benjamin's lamentably unfinished but fine analysis. Benjamin speaks of Baudelaire's "empathy with inorganic things" as "one of his sources of inspiration."[84] Benjamin goes on to connect this sensitivity to the construction of a "universe of commodities"—the phantasmagoric world of commodity fetishism that "stands in opposition to the organic" and "prostitutes the living body to the inorganic world."[85] Benjamin argues that Baudelaire expresses, with great ambivalence, the emergence of this inorganic universe in Paris, "the capital of the nineteenth century."[86]

The mid-century transformation of Paris had the effect of turning the city into one vast interior space. Changes in illumination technology had an important role in creating this sense of enclosure: "The appearance of the street as an *intérieur* in which the phantasmagoria of the *flâneur* is concentrated is hard to separate from gaslight."[87] First in the arcades and then in the streets, the gas lanterns "made the crowds feel at home in the open streets even at night, and removed the starry sky from the ambience of the big city more reliably than this was done by its tall buildings."[88] Benjamin stresses the significance of the arcades, covered by glass, floored with marble, and lighted from above by day and by gas at night, in giving the sense of interiority to the nineteenth-century city. The arcade is the classical form of the *intérieur,* but the *flâneur* sees the entire street as an *intérieur,* just as he sees the interior aisles of the department store as a street.[89] The psychological effect of this oppressive sense of enclosure is the malady Baudelaire termed *spleen.* As T. J. Clark explains, *spleen* is "the price of artifice . . . a grim and dreary conviction that the world itself has nothing to show us, a 'gloomy incuriosity' which is bound up with the collapse of nature. *Spleen* is . . . the world's deadly mimicry of our own lassitude and despair. . . . The sky is a lid, the earth is a cell, rain imitates the bars of an enormous gaol: Nature, for the most part, is a poor imitation of the world we have made."[90]

The city, though, is itself a poor imitation of an unnatural world, being too large and sprawling to be entirely a "world we have made." The smaller the scale, the more complete can be the hermetic withdrawal from the organic. In Clark's words, "artifice and nature coincide when the former becomes totally sealed and self-enclosed."[91] Maybe an entire city cannot be "totally sealed and self-enclosed," but a building can be. A domestic interior can come close to approximating the ideal of a self-constructed, self-enclosed consumer paradise.

The English country house, for example, embodies this utopian vision. An outstanding case is Fonthill, the estate inherited by William Beckford, already mentioned as the author of a truly terrifying vision of hell: the subterranean palace of Eblis in his novel *Vathek*. That imaginary palace is a classic of subterranean sublimity; Fonthill is a classic example of pseudo-subterranean fantasy. In a note added in 1838 to a letter he had written in 1782, Beckford describes how he had deliberately sealed off Fonthill for a party to celebrate the Christmas of 1782, which coincided with his twenty-first birthday. He had hired Philip de Loutherbourg, an artist and scene designer who was a master of lighting effects, to transform the Palladian mansion into an exotic fairyland. De Loutherbourg turned one room into an Egyptian temple and another into a Turkish palace. These and other exotically decorated rooms were all bathed in an eerie glow. In Beckford's words,

Immured we were au pied de la lettre for three days following—doors and windows so strictly closed that neither common day light nor common place visitors could get in or even peep in. . . . Through these galleries—did we roam and wander—too often hand in hand. . . . Delightful indeed were these romantic wanderings—delightful the straying about this little interior world of exclusive happiness surrounded by lovely beings, in all the freshness of their early bloom, so fitted to enjoy it. . . . I still feel warmed and irradiated by the recollections of that strange, necromantic light which Loutherbourg had thrown over what absolutely appeared a realm of Fairy, or rather, perhaps, a Demon Temple deep beneath the earth set apart for tremendous mysteries. . . .[92]

This mock underworld was the paradise of the private consumer. Every detail was arranged, every sense gratified. It may not have been completely sealed off from nature (there were foods and flowers on the tables), but any element of nature permitted within was carefully arranged to produce sensual pleasure. All the unpleasant aspects of nature were kept outside. ("Whilst the wretched world without lay dark, and bleak, and howling, whilst the storm was raging against our massive walls and the snow drifting in clouds, the very air of summer seemed playing around us."[93]) The technology of the underworld paradise was not productive, but magical; more precisely, its purpose was to produce the supernatural. As in the sunken paradises of the vril-ya and Captain Nemo, light glowed as if from nowhere, music strained forth as if from nowhere, and tables laden with food and flowers "glided forth, by the aid of mechanism at stated intervals, from the richly draped, and amply curtained recesses of the enchanted precincts."[94]

Immediately after this party, Beckford left Fonthill for his London residence. In January 1782, with visions of pleasure still whirling in his brain, he sat down and composed *Vathek* (in French) in three days and two nights, not pausing even to change his clothes. In Beckford's case there seems a close connection between sublimity and beauty, nightmare and dream, the enticing terror of the imaginary palace in *Vathek* and the actual Fonthill. The guests at the party had been mostly young women and handsome boys, including William Courtenay, who had become Beckford's lover. At that "voluptuous festival" the "seductive influences" (these are Beckford's words) were as much sexual as aesthetic. Beckford's guilts and conflicts regarding this homosexual affair seem to have been transferred primarily to the fictional palace created on paper so shortly thereafter.

But it is not just Beckford who suggests the ambiguities that lurk in paradise ("a realm of Fairy, or rather, perhaps, a Demon Temple . . ."). The ambivalence so celebrated in theories of sublimity persists in the aesthetic of magic. Like a sublime spectacle, a fairytale one makes the human observer feel small and impotent. Fairylands are places where wonders are performed by invisible powers.[95] The fairytale analogy is intended as praise; however, magic, like sublimity, arouses fear. What is magic is not under human control. A technology that has automatic results is pleasing when we like the results and frightening when we do not. (This source of fright has become a staple of science-fiction and horror movies: the appliances or automobiles we normally trust turn into diabolical, self-willed, even murderous devices.) The fatalism inherent in technological sublimity—where a mine explosion can seem as inevitable as a volcanic eruption—also dominates the electronic fairyland. Whether technology overpowers us or seduces us, it still makes us feel helpless. The ambivalence so obvious in images of technological sublimity persists in images of the electronic fairyworld, though less obviously. Open fears are replaced not by confidence, but by hidden fears.[96]

Despite these ambiguities, the fantasy of the enclosed artificial environment has flourished, primarily because it is so marketable. The media room of a private home is only one contemporary example of a marketable paradise. The pseudo-subterranean, high-tech *salon* that Beckford, Bulwer-Lytton, Verne, and Tarde imagined as a refuge from reality can even more clearly be recognized in the first-class airplane cabin, the hotel suite, the limousine, the executive office, the fine restaurant, the shopping mall. In all these environments, the world has not so much been disenchanted as reenchanted. The prophets of regional decentralization, like so many other social prophets of the late nineteenth century and the early twentieth, assumed we would espouse the values of order, efficiency, and rational planning. Instead, fantasy worlds lure us on every side.

The prophets of rationality were not entirely wrong, however. What has been rationalized and planned is fantasy itself. In a very calculating way, business exploits the imagination to market not just specific goods but a whole commodity-intensive way of life, which depends on the creation of a whole series of pseudo-subterranean environments.[97] Benjamin commented on the petty bourgeoisie to which Baudelaire belonged that "inevitably, one day many of its members had to become aware of the commodity nature of their labor power."[98] Until that day arrived, however, they would be "permitted, if one may put it that way, to pass their time." During that historical interlude of whiling away time, their share would be "enjoyment, but never power."

But even enjoyment is tainted and ambiguous, because the technological environments of consumer pleasure never entirely replace the larger environments of nature and society. In those larger environments, technology intrudes rather than displaces; its damaging effects are evident and troubling. In 1833 Beckford wrote to a friend in Geneva, lamenting what he had heard about developments in the surrounding areas of Switzerland:

I wager that the beautiful changes that have taken place in the area of Vaud will displease me extremely—boring long streets, big roads, English gardens,

Parisian houses (bad copies of mediocre originals)—buildings that all look alike, and that only make, so to speak, a sort of universal city extending from one end of Europe to the other. . . .

There is no longer any Country left anywhere—they are cutting down the forests, they are raping the mountains—they only want canals—they laugh at rivers—everywhere Gas and steam—the same smell, the same billows of execrable smoke, thick and fetid—the same common and mercantile glance wherever they find themselves—a deadening monotony, and a blasphemous artifice spitting every minute in the face of Mother Nature, who soon will find her children changed into Automatons and Machines.[99]

This letter was written five years before the one in which Beckford described the Fonthill Christmas party. The artifice that is so delightful at the party—the lighting, the magically gliding tables, the music, the warmth—all this has to be contrasted with the "blasphemous artifice" of urbanization and industrialization in the Swiss Alps. These two attitudes toward artifice are not contradictory but complementary. The enclosed artifice of Fonthill (literally enclosed: the estate was surrounded by walls 12 feet high) was so attractive precisely because it offered a refuge from the unsettling intrusion of technology into the natural landscape. Within those walls Beckford could create his own highly planned version of nature—for example, a "Fairie's Lawn" that led down primrose paths to a valley of shadows, with "deep caverns of the most romantic form."[100] There natural beauty was safe from desecration by the "boring long streets" of an ever-encroaching "universal city."

The two basic types of technological environment—one invading the natural environment, the other sealed off from nature—are dialectically related. The more human-made structures degrade the natural environment, the more alluring becomes the self-enclosed, self-constructed paradise. Technological blight promotes technological fantasy. Beckford's two letters should be read together. His quest for ultimate pleasure in an enclosed environment and his despair over the ravaging of the natural environment: this particular combination of powerful emotions is a structure of feeling that does not belong to Beckford alone but to a class and an age.[101]

When Beckford wrote these letters (and this is certainly one reason for their emotional intensity), Fonthill was gone. A series of extravagant projects on the house itself and the grounds had ruined Beckford financially, and in 1822 he had sold the estate to a gunpowder manufacturer. Shortly afterward, one of its splendid towers (it was well over 200 feet high, and was modeled after the octagon of Ely Cathedral) had collapsed, leaving much of the building in ruins.[102]

In the United States the typical response to the intrusion of technological change is a pastoral retreat to the natural environment; in Europe the typical response is an aesthetic retreat to the artificial environment. Like any grand generalization, this one is both inaccurate and truthful. It is most truthful and least inaccurate when applied to the French symbolist writers of the late nineteenth century, all of whom were deeply indebted to Baudelaire and shared his ambivalence toward modernity. As the critic Jules Lemaître wrote in 1895, "[Baudelaire] curses 'progress,' he loathes the industry of the century, and yet he enjoys the special flavour which this industry has given today's life. . . . I believe the specifically Baudelairean is the constant combination of two opposite modes of reaction. . . ."[103]

This "constant combination" of revulsion against and delight in the artificial environment is the emotional structure that unites the works of Baudelaire's symbolist admirers, including Villiers de l'Isle-Adam, the Goncourt brothers, Joris-Karl Huysmans, and Marcel Proust. Proust's cork-lined, shuttered bedroom (where he slept by day and composed by night) and the Goncourt's famous house (crammed with objets d'art, no "common" object being permitted within) are real-life attempts to create an all-encompassing environment of art and comfort, one hermetically sealed from the ugliness and friction of everyday reality. In similar style, the fictional dandy des Esseintes, the hero of Huysman's 1884 novel À Rebours [Against the Grain], shuts himself up in his house at Fontenay, decorating it with exquisite taste and filling it with gadgets and art treasures ("a hermitage combined with modern comfort, an ark on dry land, nicely warmed"). Axel, the eponymous hero of Villiers's drama, and his lover Sara, commit suicide in the underground vaults of the family fortress after Axel persuades Sara that if they should leave their buried paradise alive they will be unable to endure the banality of surface life.

This theme of aesthetic enclosure, of creating a pseudo-underworld of aesthetic perfection, so dominates late-nineteenth-century French literature that to deal with it adequately would require a separate volume.[104] The one example that must be discussed, because it is such a significant and neglected work of technological imagination, is Villiers's splendid, ironic, philosophical novel *L'Ève future* (1886).[105]

Like Beckford and Bulwer-Lytton, Villiers (1838–1889) was a writer of aristocratic lineage who disdained the modern world of bankers, industry, and democracy. Villiers's contempt for modernity was intensified by the contrast between his family's past fortune and its present decay.[106] He could trace his ancestry back to the tenth century, and his forebears included the founder of the Order of the Knights of Malta. The family fortunes, however, declined and were eventually ruined by the French Revolution. Villiers's father lost anything that was left in mad speculations to raise sunken galleons and to open Peruvian gold mines.

Villiers gradually moved from Brittany to Paris, and by the 1860s he had made friends with writers there (above all with Baudelaire) and had begun to write himself. He earned a reputation in the cafés as a brilliant, if highly eccentric, storyteller. By the 1870s Villiers was being slowly smothered by debts, poverty, and family miseries, and he began to withdraw from society. After years of abject poverty and utter isolation, Villiers finally began to receive recognition as a writer in the 1880s, thanks in part to the support of his new friend Huysmans. Villiers had begun to write *L'Ève future* in 1877 or 1878, and a first draft appeared in 1880. The final version was published in 1886, by which time Villiers was broken in health. He died in 1889 from stomach cancer.

L'Ève future opens by introducing its hero: "Twenty-five leagues from New York, at the heart of a network of electric lines, is found a dwelling surrounded by deep and quite deserted gardens. . . . This is Number One Menlo Park; and here dwells Thomas Alva Edison, the man who made a prisoner of the echo."[107] In a brief prefatory note Villiers explains that he is presenting not the real Edison but the symbolic one, the legendary genius: "'The Sorcerer of Menlo Park,' and so forth—and not the engineer, Mr. Edison, our contemporary."[108] Edison was very much in the news when Villiers was writing *L'Ève future,* in the late 1870s. In the fall of 1878 the inventor began to work on incandescent lamps, and after only a week of experimentation he "announced in typically bombastic fashion that he had solved the problem of electric light."[109] The pronouncement was premature but widely heralded. In December 1879, Edison illuminated his Menlo Park laboratory and grounds with over fifty carbon-filament lamps in a well-advertised spectacle that attracted many visitors from New York.[110]

Villiers's novel begins with this not implausible scene: Thomas Alva Edison, sitting in what would now be called his media room (it contains, among other devices, a telegraph receiver, a telephone, a phonograph, an electric bell, a magnesium reflector, and electric and hydrogen lights) in the cottage at Menlo Park ("like a castle lost in the woods . . . a kingdom to itself"),[111] watching the sun set and meditating somberly on his bad luck at being born too late. He has invented a machine capable of recording any sound, but all people can say into it are stock phrases and childish jests. What a tragedy that he was not around with his phonograph when God declared "It is not good for man to live alone!" or when the Sybils chanted or the oracles spoke. "Well," he thinks, ". . . until things change I'll just have to keep secret the amazing, the ultimate development of my research . . . which I have, right here, underground. And Edison tapped lightly on the floor with his foot."[112] At this point Edison receives a telegraph message from Lord Ewald, saying that he will be arriving that evening. Edison is delighted. Ewald, an astonishingly rich and handsome English noble, had rescued him some years before when he had been dying of hunger and had collapsed in the streets near Boston.

When Ewald arrives, Edison immediately discerns that this time it is the noble who needs help. Once persuaded by Edison to confess the source of his evident misery, Ewald explains that he has had the misfortune to fall in love with one Alicia Clary, a "bourgeois goddess"[113] whose ravishing, utterly perfect body holds a soul utterly barren and mediocre. The disparity is so maddening that Ewald intends to kill himself that very evening. "Ah!" he cries, "Who will deliver this soul out of this body for me?"[114]

Edison will. In order to save Ewald, he reveals the goal of his underground research: the creation of an artificial, electromagnetic woman. Edison seals off the already remote cottage, closing the windows, bolting the door, and turning off all the communication channels. Then he pushes a switch that brings up from below, with a roar as if from an earthquake, a platform on which stands, veiled and shadowed, the android Hadaly. Edison proposes to endow this half-formed creature with the perfect body of Alicia Clary and with a soul worthy of Ewald. He speaks like the magic sorcerer in a fairytale: "My lord, . . . I should warn you that we are now going to depart together from the domains of everyday life. . . . Indeed, we are going to leave the realm of Life properly so called, and penetrate another world of phenomena which will surprise and even astound you."[115]

Ewald yields to temptation and seals the pact with Edison. (Villiers's language is full of references to the myths of Faust and Prometheus.) Together they will, in Edison's words, "venture into the ARTIFICIAL and its untasted delights."[116] They put on bearskin coats, light cigars, and descend into Edison's underground laboratory, which has been fashioned out of two enormous caves where the Algonquins had once buried their dead. Here Hadaly dwells in a subterranean realm of magic. "It's a little like fairyland, this kingdom of hers," explains Edison. "Everything works by means of electricity."[117] As the two men step across a luminous threshold, the door having swung open as if on enchanted hinges, Ewald finds himself in a spacious underground chamber; it reminds him of those under the palaces of Baghdad that served to fulfill the fantasies of the caliphs.[118] In this "underground Eden" (the title of the third and central section of L'Ève future) Ewald sees a sort of Rosicrucian temple of technology: "A powerful pale blue light flooded the entire immense area. At intervals, enormous pillars supported the basalt dome. . . . At the center of the vault, dangling from a long golden wire, hung a giant lamp, blazing like a star, but with its electric rays softened by a blue shade. The vault itself, jet-black and of enormous height, loomed with the solidity of the tomb over the brilliance of this fixed star. . . ."[119] Beside one of the pillars stands Hadaly, resting her hand on a grand piano. Behind her, at the back of the room, cascades a "Niagara of flowers," while a flock of artificial birds chatter around her. The only hint of the organic is the water that rises in an elegant plume from an alabaster fountain.

Edison proceeds to explain how he manufactured each part of the android so that, unlike the clumsy automatons of earlier artisans, it would perfectly mimic the exterior appearance of living reality. Ewald is amazed at the wonders this shaman-scientist has performed:

—*You have here a sort of scientific materialism that puts to shame the imaginary world of the Arabian Nights! cried Lord Ewald.*
—*But you must also realize, Edison replied, what a marvelous Scheherazade I have here in Electricity. ELECTRICITY, my lord!*[120]

As Edison dissects Hadaly, so to speak, showing Ewald how every feature has been constructed, Ewald becomes even more incredulous. Edison responds that "the problems of creating an electro-magnetic being were easy to solve, *the result alone was mysterious.*"[121]

After everything has been explained (with many if not very convincing technical details), Edison and Ewald ascend from the "magic tomb" of Hadaly and return to "the land of the living."[122] Alicia Clary is summoned to Menlo Park on a pretext, so that Edison can precisely imprint her every feature and gesture onto the android. After three weeks, Hadaly's body has been recast in Alicia's shape. Her soul is provided through the medium of a hypnotized woman. The results are so convincing that Ewald cannot tell the difference in appearance between the living woman and the android. Enchanted by Hadaly, whose soul is as lovely as her body, he concludes that the dream woman is far superior to the natural one.

But in the end the power of nature—or rather of the divine— proves superior to the power of artifice. Hadaly is stored in a magnificent coffin-like chest so she can be carried back to Ewald's English castle on the ocean liner *Wonderful*. Three weeks after Ewald's departure, Edison reads in the newspaper that the *Wonderful* has been lost at sea. A fire that started in the cargo compartment spread quickly to the rest of the vessel. Lifeboats were launched. At this juncture, strange to say (the news story reports), a young English lord tried to rush into the burning cargo compartment; in a frenzy, he offered a stupendous reward to anyone who would save a chest stored there. Eventually he was subdued and was carried unconscious into a lifeboat. Another lifeboat, carrying Alicia Clary, capsized and sank. A bell rings in Edison's study, and a telegraph message comes:

My friend, only the loss of Hadaly leaves me inconsolable—I grieve only for that shade. Farewell.

—Lord Ewald

Edison sinks into a chair and looks out at the moon, lifting his eyes as he ponders the fate of his neotechnic Eve. Fantasy fades. He returns to the sublime contemplation of darkness and infinity:

. . . he listened to the indifferent winds of winter, whistling and howling through the bare branches—then, raising his eyes even higher toward the luminous sphere which still shone, unmoved, through the gaps in the heavy clouds, and sent their glints forever through the infinite, inconceivable mystery of the heavens, he shivered—no doubt, from the cold—in utter silence.[123]

Degeneration and Defiance in Subterranean Society

In the words of one admirer, "Villiers seems to be a man of yesterday, yet everything he writes is soaked with the premonition of tomorrow."[1] A fervent if unorthodox Catholic, fascinated by occultism, Villiers at once looks back to the mysticism of Poe (to whom he was often compared) and forward to the technicism of Wells. If Villiers seems to glorify the wonders of technology, he does so only to demonstrate their inferiority to those of God. If the seeming miracles of technology can be accepted on faith, he argues, then so can the true miracles of divinity. No one describes more splendidly the allure of the neotechnic paradise—and no one judges it more sternly as false, fallen, and pathetically vulnerable. *Ève* was written for what Villiers considered a generation dwelling in a flat universe of material reality, a generation forgetful of realities that are vertical and invisible. It is a profoundly religious work.

Those who knew Villiers commented repeatedly on his otherworldliness, which lay somewhere between mysticism and madness. He gave the impression of living aloof from his own suffering and humiliation. It might just as well be said, however, that Villiers lived on a level below that of ordinary humanity. He is a prime example of what Irving Howe has called "the underground man," a literary figure and a social type that entered European consciousness in the nineteenth century: "A creature of the city, he has no fixed place among the social classes; he lives in holes and crevices, burrowing beneath the visible structure of society. . . . Even while tormenting himself with reflections upon his own insignificance, the underground man hates still more—hates more than his own hateful self—the world above ground."[2] No other writer discussed here is less at home in the world of the second industrial revolution than is Villiers. And so it is especially striking to see how many other writers, far more reconciled with their times, echo the fundamental theme of *Ève*: that the technological paradise, however tempting, is false and potentially fatal. Bulwer-Lytton, Tarde, Forster, and Wells repeat this warning in their subterranean narratives. They repeat it, however, in a secularized version. Their concern is less with the danger of transgressing divine laws than with the peril of contravening natural ones. What Villiers considers sinful man's fall from grace, they call degeneration.

Although the terms *degeneration* and *decadence* are typical of the nineteenth century, they express the primitive fear that mastery over nature brings divine retribution. This fear was expressed in the myth of Prometheus, and it inspired the rituals of appeasement that preceded mining as late as the Middle Ages. Not just mining but technological achievement in general has traditionally aroused such dread.[3] In the secular literature of the West, this anxiety about over-civilization has often been expressed in pastoral language. Not only Virgil but countless other Greek and Latin writers sang the praises of husbandry and warned against the corruption of city life. Later on, Elizabethan poets contrasted the innocent happiness of shepherds with the false pleasures of courtiers. Still later, Jean-Jacques Rousseau, in his famous 1750 response to the query of the Dijon Academy, argued that the arts and sciences of civilization corrupted rather than purified morals. Even John Stuart Mill, usually considered a prime spokesman for nineteenth-century civilization, argued that its advantages had been bought at a high price: ". . . the relaxation of individual energy and courage; the loss of proud and self-relying independence; the slavery of so large a portion of mankind to artificial wants; their effeminate shrinkage from even the shadow of pain; the dull unexciting monotony of their lives, and the passionless insipidity, and absence of any marked individuality, in their characters. . . ."[4]

In the middle of the nineteenth century these ancient anxieties came to be articulated in the prevailing language of biological materialism. As early as the 1840s the term *degeneration* was being used to warn of various physical, mental, and moral maladies. In 1857 the

French physician Bénédict-Augustin Morel, who had a particular interest in mental illness, published his highly influential *Traité des dégénérescences physiques, intellectuelles et morales de l'espèce humaine.* Morel stressed the impact of heredity and, in a later textbook, proposed that "each new generation received a heavier and more destructive dose of whatever the evil influence was."[5] Morel's largely hereditary theory merged almost immediately with the emphasis on environmental factors introduced by Charles Darwin and (to a lesser extent) by Herbert Spencer.[6] According to Darwinist and Spencerian principles, biological evolution originates in the general struggle for existence. Therefore, it seemed logical to conclude that the harsher the environment, the fiercer would be the struggle and the more rapidly evolution would proceed. In an age when evolution was habitually equated with progress, the virtue of a challenging environment seemed self-evident. If conditions were too tame, there would be little struggle and therefore little progress. Thomas Huxley warned: "If we may permit ourselves a larger hope of abatement of the essential evil of the world . . . I deem it an essential condition of the realization of that hope that we cast aside the notion that escape from pain and sorrow is the proper object of life."[7]

The danger of a highly technological environment, then, was that it might succeed too well in insulating humankind from nature's hazards. If a nation or a race succeeded in creating an artificial environment of ease, comfort, and efficiency, it would be doomed by its own technological success. The inevitable result would be flabbiness—mental, moral, physical—and eventual submission either to inner decay or to external conquest. In this way, too, the idea of progress, the ruling concept of the century, was mingled with visions of ruin. Once again, it seemed that progress inevitably brought destruction—not only visible damage to landforms and settlements, but also invisible damage to human character.

Of all the writers of underground narratives, H. G. Wells most consciously used his stories to consider the relationship between technological progress and human degeneration. Wells escaped from a lower-class background (his mother worked as a maid, his father was an unsuccessful shopkeeper, and he was destined for apprenticeship to a draper or druggist) when he showed unusual ability in science. At age eighteen he won a scholarship to the Normal School of Science, in South Kensington, to train as a teacher. The morning he arrived, in September 1884, was, Wells later said, "one of the great days of my life." The year he attended Thomas Huxley's lectures in zoology and biology (1886–1887) "was beyond question the most educational year of my life." Huxley, who had worn himself out campaigning for science education and for the ideas of Darwin, became Wells's image of the heroic man of science. Wells was deeply impressed by his teacher's anxiety that biological evolution, far from leading to human progress, might well lead to human retrogression, since fitness to environment bore no relation to ethical principles. Wells also came to share Huxley's conviction (stated decisively in his 1893 book *Evolution and Ethics*) that the only hope for humankind was to consciously oppose natural processes in favor of self-generated ethical purposes.[8]

Wells's scientific interests were far-ranging and genuine; between the late 1880s and the late 1890s he published more than 200 scientific papers.[9] At the same time, though, he was beginning work on the scientific romances that would make his literary reputation. In 1887, the year he completed his course with Huxley, Wells started work on a fantasy called "The Chronic Argonauts," which was serialized in three parts in *Science Schools Journey* in the spring of 1888. It was first draft of *The Time Machine,* published in book form in 1895. *The Time Machine,* which brilliantly launched Wells's writing career, is still an unforgettable vision of human degeneration in a technological world.

Since evolutionary processes proceed in deep time, the narrator of *The Time Machine* must find a way to move across millions of years. The machine that allows him to do this is a bicycle-like contraption with a framework of glittering brass and nickel that carries the Time Traveler to the year 802701. There he finds what seems to be a pastoral utopia inhabited by the beautiful, graceful Eloi. There are no insects or weeds. Fruit grows everywhere. Disease has been eradicated, and everyone seems to be fed and clothed without toil. The whole earth has "become a garden."[10]

The Time Traveler is troubled, however, by what is absent from this social paradise. The Eloi have a childlike charm as they giggle, play, and chatter, but they are incapable of any intellectual or physical effort. Their only arts are dancing and singing in the sunlight. They have no curiosity, no energy. Their cities lie in ruins. The Time Traveler concludes that they represent the degeneration that followed the "too perfect triumph of man": "Strength is the outcome of need: security sets a premium on feebleness. The work of ameliorating the conditions of life—the true civilising process that makes life more and more secure—had gone steadily on to a climax. One triumph of a united humanity over Nature had followed another."[11] The conquest of nature, that long-sought goal, had ended in the destruction of human intelligence and vigor. "For after the battle comes Quiet. . . .the reaction of the altered conditions."[12]

As the narrative proceeds, however, the Time Traveler learns that his theory is incomplete. The Eloi do not live in security. They have lost the strength to resist their enemies, the Morlocks, a race of white, apelike underground dwellers—"bleached, obscene, nocturnal Things[s]."[13] The Time Traveler becomes aware of the Morlocks' existence when they steal the Time Machine. After other encounters, he realizes that the Morlocks operate subterranean machines. They feed and clothe the Eloi out of habit—and on moonless nights they prey on them for meat.

So there are two types of degeneration in *The Time Machine:* the degeneration into weakness of the overprivileged Eloi and the degeneration into brutality of the exploited Morlocks. Wells's story is therefore a good jumping-off point for this chapter and the next, for it summarizes the two major fears of decadence that might result from a technological environment. The first fear is that humanity might get its wish and devise marvelous technologies that conquer nature and fulfill every conceivable need. What if this fond dream were fulfilled? Might this supposed paradise destroy the sources of human integrity? As Irving Howe writes, "Not progress denied but progress realized, is the nightmare haunting the antiutopian novel."[14] The second fear is that such technological progress will not be realized. If humanity must continue to labor to satisfy its needs, what will happen then? Will workers become increasingly brutalized as their labor becomes more mechanized and repetitive? Will the split between labor and capital become unbridgeable? The technological trap snaps shut. Either by conquering nature or by not conquering it, humanity will degenerate.

Is there no escape? In this chapter we shall examine this question by looking at some subterranean tales based on the first assumption. Bulwer-Lytton's *The Coming Race,* Forster's "The Machine Stops," and Tarde's *Underground Man* all assume that humankind has created for itself a painless, generous technological environment. In *The Time Machine,* only the Morlocks lived underground; the feeble Eloi dwell on the surface. In these three tales, however, the paradise of ease and plenty is set below the ground, far removed from the harshness and unpredictability of nature. In the next chapter we will consider underworld realms that are variations on the theme of the Morlocks. These are found in subterranean narratives by Jules Verne and H. G. Wells in which the continued existence of a submerged working class and the continued struggle against nature are assumed.

In both types of story, the fear of degeneration is fundamentally a political anxiety. As we shall see in this chapter, the vision of an all-too-perfect conquest of nature is also a vision of the conquest of the individual by the collectivity. In the fictions of Verne and Wells, where society is not homogeneous and labor remains a fact of life, a primary anxiety is that the state will become all-powerful. In brief, human life in a highly technological environment seems to demand a high degree of authority. Degeneration is discussed in the language of biology and morality, but it expresses social and political concerns.

This is most evident in the works of Wells, for whom (as Herbert Sussman notes) "Darwinism and Marxism coincide."[15] During his years at South Kensington, Wells became attracted to socialism; he was a member of the Fabian Society between 1903 and 1908.[16] *The Time Machine* projects the contemporary class struggle into an even more radical split between brute labor and trivial consumption, between toilers and enjoyers. But Wells is hardly alone in the way he uses biological language to express political convictions. The vocabulary of degeneration (or decadence) was repeatedly used to express as a general concern about human nature what was in fact a specific fear of particular groups—a race, a nation, a gender, a class. As with other such theories of biological determinism, race, sex, and class served as surrogates for one another.[17] These groups were interchangeably described as an alien, subhuman species. It is not just that biological ideas were extended to social and political situations in which (as Huxley would point out) the analogies were false; it is also the case that the biological ideas themselves "had a social component before there was any question of reapplying them to social and political theory."[18]

Today the language of degeneracy makes us uncomfortable, as it should. We have seen what crimes can be committed in its name.

On the other hand (and this is why these chapters are being written), the concept of degeneration provides a way to consider the interactive evolution of technological and social systems. To use the formulation of Marx (who, of course, greatly admired Darwin), "production not only creates an object for the subject but also a subject for the object."[19] The language of degeneration permits us to look at the relationship between technological object and human subject without falling back upon another intellectual model associated with Marxism: the static and unidirectional model of base and superstructure. Instead, both technologies and "human nature" (understood as "social nature") are apprehended as interactively evolving systems.

From this evolutionary perspective, technologies are best considered as environments rather than as objects—an approach that is of course fundamental to this book. From such an environmental perspective, technological change is best evaluated in terms of the general direction of change rather than in terms of the supposed effects of this or that device. Technologies cannot simply be thought of as neutral tools that are used by people who are good, or bad, or some mixture of these stable elements, and that are used for ends that are good, bad, or some mixture. The user changes as the technological environment does. Both tools and users are malleable, but not infinitely so. We can assume the presence of some indestructible strivings, which may nonetheless be suppressed at any historical moment, just as we may assume unexplored potentialities.[20] The range of possibilities is large and flexible, if not unlimited.

Let us now look at some tales of possibility.

At first *The Coming Race* seems a straightforward celebration of technological progress inevitably leading to social progress. Its premise is strict technological determinism. Once the underground race discovers vril, the all-purpose source of energy, its evolutionary direction changes radically. Force disappears, since it is impossible to resist this ultimate weapon. Competition disappears, since everyone can have anything he wants. The narrator of *The Coming Race* can actually see the evolutionary change when he is taken to view a portrait gallery of the vril-ya that begins with the remotest origins of the race, some six or seven thousand years earlier. The first portraits "resembled our own upper world and European types of countenance"—faces furrowed by ambition, care, or grief. About a thousand years after the vril revolution, however, the faces become, "with each generation, more serene, and in that serenity more terribly distinct from the faces of labouring and sinful men."[21]

The argument of *The Coming Race* is that this social evolution, instead of leading to a higher civilization, is in fact a degenerative process. In a letter to his son, Bulwer-Lytton explains that his aim in writing the book is to imagine what society would be like if all utopian dreams were achieved. The result, he predicts, would be "firstly, a race . . . fatal to ourselves; our society could not amalgamate with it; it would be deadly to us, not from its vices but its virtue. Secondly, the realization of these ideas would produce a society which we should find extremely dull, and in which the current equality would prohibit all greatness."[22]

The vril-ya have gained serenity and harmony, but they have lost theology, history, art, and heroism. The uncomfortable urges that lead to great achievements as well as to great suffering have been snuffed out. Like Wells's Eloi, the vril-ya have settled down to what the narrator considers boring quiescence. The difference is that the vril-ya, who command their own technology rather than depending upon a laboring class, remain physically and morally strong. The barbarian races that still exist in other parts of the underworld pose no threat, for they can be exterminated with a whiff of vril. Technological superiority has led to moral and physical superiority—but a superiority that seems unreal, unnatural, and, to the narrator, frightening. By being exalted above the human condition of effort and worry, the vril-ya have become inhuman. When the narrator, at the end of his tale, sees the chief magistrate advancing toward him, he is terror-stricken: "On the brow, in those eyes, there was that same indefinable something which marked the being of a race fatal to our own—that strange expression of serene exemption from the common cares and passions, of conscious superior power, compassionate and inflexible as that of a judge who pronounces doom."[23] The book ends with a warning of this "people calmly developing, in regions excluded from our sight and deemed uninhabitable by our sages, powers surpassing our most disciplined modes of force." The narrator breathes a devout prayer "that ages may yet elapse before there emerge into sunlight our inevitable destroyers."[24]

Raymond Williams has said of *The Coming Race* that "its desire is tinged with awe and indeed with fear."[25] The desire arises because the vril-ya seem superhuman, the fear because they seem inhuman. They are the forerunners of the alien invaders of twentieth-century science fiction who, according to Susan Sontag, would impose a "regime of emotionlessness, of impersonality, of regimentation" on the earth.[26] Like such aliens, the vril-ya represent "the wave of the future, man in his next stage of development."[27] Bulwer-Lytton himself, of course, was quite unaware that he was sketching out a major theme of a yet-undeveloped fictional form. What then was the source of his prophetic anxiety? Who are these aliens from inner space whose command of technology threatens to give them soulless superiority over old-fashioned, muddling, imperfect humanity?

In a prefatory note to Bulwer-Lytton's *The Parisians* his son, the poet-novelist Owen Meredith, states that three of his father's last works[28]—*The Coming Race* (1871), *The Parisians* (1873), and *Kenelm Chillingly* (1873)—have the moral purpose of (generally) criticizing modern ideas and (more specifically) criticizing Darwinism, democracy, and materialism. *The Parisians* shows the ill effects of modern ideas upon an entire community; when it was published, the Paris Commune of 1871 had recently been crushed.[29] *Kenelm Chillingly* shows the ill effects of modern ideas on the individual (the hero imbibes advanced ideas and learns the hard way to value love and domestic contentment). Like *The Coming Race,* these novels express the fear that the political dreams of civilization might instead become its nightmare.[30]

Bulwer-Lytton belonged to an aristocratic age, and when he wrote these books that age was drawing to a close. In his younger days he had been a reform-minded aristocrat. Entering Parliament in 1831, he had joined with a group of philosophical reformers on the Whig left. After withdrawing from the House of Commons in 1841, however, Bulwer-Lytton returned there in 1852 as a conservative supporter of the Disraeli's Tory Democracy. There are many reasons for this conversion (his inheritance of the family property, his friendship with Disraeli, his disdain for the laissez-faire doctrines of middle-class manufacturers), but overarching all of them was a dread of the economic and political changes implied by "modern ideas." *The Coming Race,* then, presents a defense of an established social order in the guise of a defense of old-fashioned "human" virtues. What might look like progress, warns Bulwer-Lytton, is really degeneration.

Besides these fears, *The Coming Race* expresses another that is at least as powerful: the fear of female dominance. Women's liberation is an important part of the social revolution brought about by the vril revolution. For reasons both hereditary and environmental, the female vril-ya are taller and ampler than the males, and, even more crucial, they are better able to control the potent vril. As a result, "among this people there can be no doubt about the right of women."[31] The women tend to be tactful about using their powers, but they do insist upon the right to be the pursuers in courtship. Being so persuasive and powerful, the females, once set upon winning a man, "are pretty sure to run down his neck into what we call 'the fatal noose.'"[32]

The plot of *The Coming Race* springs from the narrator's discovery that he is being pursued by Zee, one of the tallest, smartest, kindest, and most determined females. He may be flattered, but he is above all terrified. Her perfection, far from being enticing, unnerves him: ". . . never in the upper world have I seen a face so grand and so faultless, but her devotion to the severer studies had given to her countenance an expression of abstract thought which rendered it somewhat stern when in repose; and such sternness became formidable when observed in connection with her ample shoulders and lofty stature."[33] The narrator is also frightened by her attentions because he knows the vril-ya strictly forbid intermarriage with another race. If he should yield to Zee's wooing, he will be reduced to a cinder by a bolt of vril. He becomes even more panicky when he realizes that Zee's parents have no control over her and that they will do nothing to interfere with her infatuation. Zee herself, knowing that her choice is frowned upon by the community, suggests that they flee to other subterranean lands or to the surface, where they can live together platonically. But a sexless relationship with a formidable Amazon is not the narrator's idea of living happily ever after; he refuses as politely as possible.

When the daughter of the chief magistrate (or Tur, as he is called) also begins to court the narrator, he is in imminent danger of death. The Tur, disgusted by his daughter's preference, orders his young son to use his vril staff to get rid of the troublesome human. Fortunately, both the Tur's son and Zee are so fond of the narrator that they disobey the law of the community. The Tur's son warns him instead of slaying him, and Zee finally uses the power of vril and her giant wings to fly him back to a mineshaft, where he is rescued.

This plot represents more than unenlightened nineteenth-century masculine anxieties. In the last paragraph, the narrator, remembering Zee's perfection, sighs that he has since been "somewhat disappointed, as most men are, in matters connected with household love and domestic life."[34] This remark severely understates Bulwer-Lytton's own spectacularly miserable domestic life. His mother appears to have been at once doting and tyrannical. When Bulwer-Lytton fell in love with a beautiful but tactless and egotistical woman, his mother warned him never to marry her—which of course made him all the more determined to do so. His mother refused to come to the wedding, and cut off his income. Bulwer-Lytton had to write frantically to provide for his family. His overwork, which aggravated his bad temper, was one factor in the breakdown of his marriage, which ended in a separation in 1836.

Lady Lytton lived abroad for a while, but in 1847 she returned to England and began to harass her estranged husband with obscene letters and various threats. In 1858 she mounted a platform where he was addressing constituents and told the crowd that her ex-husband, then serving as secretary of state for the colonies, should himself be transported to the colonies as a felon. Bulwer-Lytton swiftly arranged to have her detained as a person of unsound mind. When the press got wind of this, he was severely criticized. The government in which he was serving feared an even more sensational scandal. Finally, arrangements were made for their son to take Lady Lytton abroad. After several months she returned, and she "maintained an uneasy peace with her ex-husband for the remainder of their lives."[35]

Reading this sad story a century later, one is inclined to judge Bulwer-Lytton harshly, as a temperamental and self-absorbed husband who, by cruelly neglecting his wife and depriving her of affection or sympathy, drove her first to infidelity and then to unusual but hardly irrational personal attacks. One can also understand, however, that for him the idea of liberating women from parental and male authority touched upon a personal torment. Bulwer-Lytton's anxiety was precisely that technological progress *would* lead to social progress of the sort that was conventionally anticipated by those with modern ideas—feminism, socialism, democracy, and the like. The conquest of nature might put social power into the hands of those he considered naturally subordinate. In the topsyturvy underworld of *The Coming Race,* human weakness is a virtue and unnatural strength is degeneration.

E. M. Forster's story "The Machine Stops," written in 1909 (a little less than 40 years after *The Coming Race*), is also based on the premise that a marvelous new technology removes the need for labor. "The Machine," like vril, serves all and exploits none: "Night and day, wind and storm, tide and earthquake, impeded man no longer. He had harnessed Leviathan. All the old literature, with its praise of Nature, and its fear of Nature, rang false as the prattle of a child."[36] Forster shows us particular extensions of the Machine (medical devices plugged into the ceiling, speaking tubes, moving chairs, the wormlike Mending Apparatus), but not the Machine itself. Its invisibility makes it all the more oppressive. To today's reader, the Machine inevitably suggests a sort of supercomputer.

As a political force, too, the Machine is both omnipresent and invisible. Instead of a clearly defined state, there is an unseen Central Committee of the Machine. Although the Committee announces new policies, "they were no more the cause of them than were the kings of the imperialistic period the cause of war. Rather did they yield to some invincible pressure, which came no one knew whither, and which, when gratified, was succeeded by some new pressure equally invincible. To such a state of affairs it is convenient to give the name of progress."[37] Both technology and politics are disembodied forces, literally unseen and intellectually baffling. Things just happen.

"The Machine Stops," like *The Coming Race,* demonstrates the inverse relationship between technological advance and human decline. To do so, Forster used the by-then common conceit of treating the Machine as an evolving organism.[38] It generates its own foodtubes and medicine-tubes and music-tubes, becoming ever more complex and more aggressive. In proportion to the growing strength of the Machine, people degenerate.

There is nothing paradoxical about Forster's definition of degeneration. People literally become ever weaker, more passive, more helpless. The prime example is Vashti, one of the two protagonists of the tale: "a swaddled lump of flesh—a woman, about five feet high, with a face as white as a fungus."[39] She spends almost all her life in her hexagonal room, rolling around in her movable chair, manipulating buttons and switches, sheltering herself from any disturbance. Above all, she avoids direct sensory experience. Always on the prowl for "new ideas," she listens to and delivers lectures on topics like "Music during the Australian Period" or "The Sea" while avoiding the music and the sea. She is happy to talk with people through communications links, but she dreads even getting near them, not to mention speaking directly to them or (worse yet) touching them.

As much as Bulwer-Lytton, Forster rails against "modern ideas"; however, Forster objects less to any particular set of political or social ideas than to the general preference for abstract ideas over the direct experience of nature.[40] Forster assumes that physical and intellectual life are inseparable—that if the body deteriorates, so does the mind. The "chief sin" of his subterranean civilization is "the sin against the body . . . the muscles and the nerves, and those five portals by which we can alone apprehend." Allowed to turn to "white pap," the body has become "the home of ideas as colourless, last sloshy stirrings of a spirit that had grasped the stars."[41] The life of the machine is the enemy of human life. As Vashti's son Kuno exclaims, addressing her and us alike: "Cannot you see, cannot all you lecturers see, that it is we that are dying, and that down here the only thing that really lives is the Machine? We created the Machine, to do our will, but we cannot make it do our will now. . . . The Machine develops—but not on our lines. The Machine proceeds—but not to our goal."[42]

Kuno is the rebel-hero of the story. Despite a eugenics policy that decrees death for any child who promised to grow up physically strong ("Man must be adapted to his surroundings, must he not?"),[43] Kuno is "possessed of a certain physical strength," which he has developed by doing calisthenics and by walking and climbing through the maze of underground passages. Furthermore, he is possessed of an atavistic curiosity, particularly to see what is on the surface. Eventually, with great effort, he climbs through a dark shaft leading from a railroad tunnel, opens a pneumatic stopper that keeps the outer air from the underworld, and emerges above ground, bleeding, with a tremendous roaring in his ears and with the bitter outer air painfully filling his lungs. (Forster's description of the journey is heavy with images of childbirth.[44]) Kuno, reborn, finds himself in a grass-filled hollow in Wessex, where he gazes at the low hills and watches the sun set in the mist. Then (much as in Wells's story, where the Time Machine is stolen by the Morlocks), Kuno's respirator is snatched by the wormlike Mending Apparatus. After denuding all the living things in the dell, the Mending Apparatus curls its tentacles around Kuno and drags him back down to his room, where he awakens in "artificial air, artificial light, artificial peace."[45]

But Kuno has learned that people still live on the surface. Vashti cannot believe Kuno's report. After the Great Rebellion, she reminds him, the defeated were forced to the surface, where their bones still lie. "And so with the Homeless of our own day. The surface of the earth supports life no longer."[46] She shakes her head with pitying wonder. Obviously her son is mad, and obviously he will end in Homelessness too.

But she is wrong. Since the Machine is an evolving organism, it can degenerate too. It begins to break down just as mysteriously and inexorably as it had progressed. First, instead of soothing music, jarring noises invade the hexagonal cells—"And so with the mouldy artificial fruit, so with the bath water that began to stink, so with the defective rhymes that the poetry machine had taken to emit. All were bitterly complained of at first, and then acquiesced in and forgotten. Things went from bad to worse unchallenged."[47] Then the sleeping apparatus breaks down. Next it is announced that the Mending Apparatus itself needs repairs. As the subterranean atmosphere grows dirtier and the communication system breaks down, a terrible silence replaces the hum of the Machine.

The final crisis is like an earthquake. Everything cracks and rumbles, the floor heaves, darkness falls, and "the original void returned to the cavern from which it had been so long excluded."[48] At the last moment, Vashti hears Kuno's voice. Weeping, they embrace. Just before an airship crashes into the subterranean honeycomb, destroying everyone and everything, he reminds her that although the subterraneans will die, the Homeless "are hiding in the mist and the ferns until our civilization stops." "Humanity," he says, "has learnt its lesson."[49]

The lesson is that a technological environment is fatal to humanity. Both Forster and Bulwer-Lytton argue that degeneration results when technological development reaches a point where human beings detach themselves from the natural order of things. They both stress that technological development destroys what they consider the natural sexual order, either by masculinizing women into formidable scholar-suitors or by feminizing men into pale weaklings. For Bulwer-Lytton, however, the natural order is essentially a social one, defined in terms of aristocracy and male dominance. Forster thinks of the natural order in the more literal terms of the physical landscape and the bodily senses. In both cases, though, when technology rather than nature becomes the environment of life, human evolution moves in a disastrous direction, toward a loss of autonomy so profound that people are no longer aware of it.

The "lesson" of Gabriel Tarde's *Underground Man,* it seems, could not be more different. In the subterranean realm of Tarde's neo-Troglodytes, a technological environment produces a flourishing of art, individuality, and sociability. This is a genuine utopia, not a fake one. But when we look more closely, we see that the achievement of this utopia has nothing to do with technological invention. The great difference separating Bulwer-Lytton and Forster from Tarde is that whereas the Englishmen portray what happens when a technological revolution (the discovery of vril, or the universal establishment of the Machine) directs social change, Tarde depicts what might happen if an authentic social revolution were to direct technological change.

Although Tarde is far less known today than his competitor Émile Durkheim, he was an eminent sociologist in the same period. He was best known for his writings on criminology and economics.[50] Like Wells, Tarde admired the scientific approach and consciously used his subterranean fantasy as a narrative thought experiment. In particular, *Underground Man* uses the experimental technique of radical "world-reduction." The neo-Troglodyte narrator explains:

[*Our civilization*] *consists in the complete elimination of living nature, whether animal or vegetable, man only excepted. That has produced, so to say, a purification of society. Secluded thus from every influence of the natural milieu into which it was hitherto plunged and confined, the social milieu was for the first time able to reveal and display its true virtues. . . . It might be said that destiny had desired to make in our case an extended sociological experiment for its own edification by placing us in such extraor-dinarily unique conditions.*[51]

The experiment demonstrates that terrestrial nature had been "an unsuspected drag . . . on the progress of humanity."[52] Moral, intellectual, and aesthetic life all flower once humanity is freed from the degrading need to deal with nature's caprices and hardships. The underground people treat one another with "an indescribable courtesy,"[53] of which the perennial and unseen source is love—"the true religion, universal and enduring, that pure and austere moral which is indistinguishable from art."[54] In science, architecture, philosophy, painting, and literature, the underground dwellers find endless delight and intellectual pleasure in their artificial environment. Astronomers debate endlessly about the stars they never see, and artists sketch animals and landscapes more lovely than those that actually existed on the surface. The imperfections of nature are replaced by the perfections of culture.

According to biographer Terry Clark, Tarde, like Baudelaire, believed that "artifice . . . is nature's final fruit" and that "nature is disclosed . . . only when artifice is brought to the pitch of perfection."[55] For the sociologist Tarde, society was the highest art form. He was a social aesthete. The crucial point about Tarde's thought experiment—the reason why love and art blossom in his underworld—is that the natural environment has been replaced not by a technological environment but by a social one. What constitutes (in Tarde's words) a "truly social revolution" is the collective decision to renounce utilitarian activities in favor of aesthetic ones: "The ancient social ideal was to seek amusement or self-satisfaction apart and to render mutual service. For this we substitute the following: to be one's own servant and mutually to delight one another."[56]

In Tarde's *Underground Man,* the change in ideals comes about slowly and painfully. The first step is taken in response to natural necessity: the waning of the sun has reduced humankind to a chilled and dejected remnant. The hero-genius Miltiades, through the force of his charismatic personality, inspires his listeners to move underground, urging them to take not, as Noah did, "this miscellany of living contradictions which for so long was so foolishly worshipped under the name of Nature," but instead "all artistic and poetic beauties . . . our real treasures, our real seed for future harvests."[57]

But the social revolution is not yet complete. Once humanity has moved below the surface, a terrible war, "hideous and odious, revolting beyond all expression,"[58] breaks out between the centralist, industrial cities and the federalist, artistic cities. Only when the industrial cities are defeated does society unite in its resolution to produce art rather than to consume goods. "The quota of absolute necessities being thus reduced to almost nothing, the quota of superfluities has been able to be extended to almost everything. Since we live on so little, there is abundant time for thought. A minimum of utilitarian work and a maximum of aesthetic, is surely civilization itself in its most essential element."[59] After the defeat of the industrial cities, technological choices are guided by these social principles. The neo-Troglodytes discover beautiful grottoes, which they extend to make their dwellings. They melt ice for water, and they get meat from animal carcasses frozen in glaciers. They tap the central heat of the earth to generate power. Scattered centers of fire produce free electricity for lighting. Like subterranean nut and berry pickers, the neo-Troglodytes live off the bounty of subsurface nature. Like Virgilian shepherds, they combine material simplicity with spiritual elevation. The shortages and disasters of nature are banished, but not its abundances and its beauties. Tarde's book is an underworld pastoral.

Still, there are rebels. The main reason for their discontent is the sexual code that is one facet of the social revolution. The war between the industrial and the artistic cities was followed by another "still more bloodthirsty conflict" between "free-thinking cities" advocating sexual freedom and "cellular cities" fighting for "prudent regulation."[60] The victory of the latter was necessary, explains the narrator, because with limited food supplies it was "a duty for us rigorously to guard against a possible excess in our population." "We have," he continues, "been obliged to forbid in general under the most severe penalties a practice which apparently was very common and indulged in *ad libitum* by our forefathers."[61] So, while love is everywhere, sex is strictly regulated by informal but powerful community pressure. The right to have children—"the monopoly and supreme recompense of genius"[62]—is granted to a couple only when the man has produced an artistic masterpiece under the inspiration of the woman. A first offense against this rule is punished by degradation, and a second by throwing the offender into a lake of petroleum. These penalties have become unnecessary, however, because the force of public opinion is so strong. Some "saintly souls" are able to sublimate their sexual desires into aesthetic ones, but many lovers suffer from the strain of renunciation. Indeed, quite a few go mad from love and die; others "courageously get themselves hoisted by a lift to the gaping mouth of an extinct volcano and reach the outer air which in a moment freezes them to death. They have scarcely time to regard the azure sky—a magnificent spectacle, so they say—. . . then locked in each other's arms they fall dead upon the ice! The summit of their favorite volcano is completely crowned with their corpses which are admirably preserved always in twos. . . ."[63]

Although the "prudent regulation" causes individual suffering, its wisdom becomes evident when, in the Year of Salvation 194, a bold excavator drills into an open space and discovers "a veritable underground America, quite as vast and still more curious." It proves to be "the work of a little tribe of burrowing Chinese," though Tarde's reference to America and another remark about Christopher Columbus suggest that he is really thinking of the United States. In any case, this race is unaesthetic, unencumbered by the past, and undisciplined by natural scarcity. To escape the cold, these "degraded beings" had hastily crawled underground without packing up any books or artworks. Once there, they had reproduced without limit. Not being able to feed themselves from animal carcasses, they had reverted to "ancestral cannibalism," using the millions of Chinese buried in the snow "to give full vent to their prolific instincts." The narrator concludes with disgust: "In what promiscuity, in what a slough of greed, falsehood and robbery were these unfortunates living! The words of our language refuse to depict their filth and coarseness." Some suggest exterminating these savages, while others propose making them slaves or servants, but these ideas are rejected. After a futile attempt to civilize "these poor cousins," "the partition [is] carefully blocked up."[64]

The crucial variable in this underground thought experiment is neither technology nor nature but society. The "Chinese" too dwell underground, but the outcome of their subterranean evolution has been degeneration, not perfection. The reason is that they have not undergone a "truly social revolution" in which aesthetic values are given precedence over economic ones. For Tarde, the opposing poles of value are not nature and technology but economics and aesthetics. Economic goals isolate whereas art and love unite. If there is an inverse relationship here, it is not between technological progress and human degeneration but between economic pursuits and genuine sociability.

Tarde too attacks modern ideas, not because they are unnatural but because they are anti-social. In *Underground Man* he speaks as a friendly critic of socialism in calling attention to the materialistic bias of contemporary socialists. Their error, he says, is that their goal of an "intense social life" is inevitably based on "the aesthetic life and the universal propagation of the religion of truth and beauty." "The latter," he continues, "assumes the drastic lopping off of numerous personal wants. Consequently in rushing, as they did, into an exaggerated development of commercial life, they were marching in the opposite direction to their own goal."[65]

The ideal society results when an authentic social revolution (that is, the limitation of material needs and of population growth) takes precedence over a technological revolution. On this point Tarde's utopia, Bulwer-Lytton's pseudo-utopia, and Forster's dystopia are in essential agreement. In the interactive evolution of humanity and technology, they all warn, degeneration is the result when technology takes command.

How valid are these forecasts? We can understand Villiers's warnings against a fascination with gadgetry that invests technologies with pseudo-divine powers; we can appreciate Bulwer-Lytton's predictions that technological advances can put power into the hands of the hitherto powerless; and we can recognize Forster's description of human helplessness when complex technologies break down, and Tarde's anxieties about unchecked population growth. But what about the dominant theme: that a too-comfortable technological environment will weaken us physically, morally, and intellectually? Such warnings of degeneration have continued into our own century. In one well-known example, in *The Road to Wigan Pier,* George Orwell explains why he is troubled by Wells's optimistic visions of the future as expressed in various technological utopias subsequent to *The Time Machine:*

All mechanical progress is towards greater and greater efficiency; ultimately, therefore, towards a world in which nothing goes wrong. But in a world in which nothing went wrong, many of the qualities which Mr. Wells regards as "godlike" would be no more valuable than the animal faculty of moving the ears. . . . In a world from which physical danger has been banished— and obviously mechanical progress tends to eliminate danger—would physical courage be likely to survive? Could it survive? . . . As for such qualities as loyalty, generosity, etc., in a world where nothing went wrong, they would be not only irrelevant but probably unimaginable. The truth is that many of the qualities we admire in human beings can only function in opposition to some kind of disaster, pain or difficulty; but the tendency of mechanical progress is to eliminate disaster, pain and difficulty.[66]

In sum, says Orwell, "what is usually called progress also entails what is usually called degeneracy."[67]

But this future in which nothing goes wrong is scarcely recognizable in the present. A half-century after Orwell, and three-quarters of a century after Wells and Forster, mechanical progress shows little sign of eliminating disaster, pain, and difficulty. Much of the world's population still confronts daily the challenges of hunger, thirst, disease, cold, or heat. Even for the more swaddled rich of the world, life in a highly technological environment hardly seems painless and stressless. Technology has not so much eliminated danger as it has substituted new dangers for old ones. If we do not face bears in the woods, we face busy intersections. Far from shunning physical and mental challenges, we use technological ingenuity to create new ones: computer games, hang gliders, boardsails. And far from living in a riskless environment, we have developed an enormous literature on technological risk. Technology has not so much replaced nature as it becomes a second nature, with its own attendant pleasures and hazards.

Once again, "The Machine Stops" seems especially prescient in its technological vision. Forster's insight was that with machines, as with nature, a lot can go wrong to cause "disaster, pain and difficulty." His description of the Machine's slow degeneration rings far more true to a contemporary reader than any predictions of human degeneration. What also rings true is the response of the nervous Vashti, the quintessential consumer under late capitalism who enjoys a high degree of comfort at the price of a chronic sense of tension and insecurity. Her dominant emotion is irritation, "a growing quality in that accelerated age."[68] Far from a world where nothing goes wrong, she lives in a world where everything is always going wrong. For the most part the annoyances are petty (poor-quality food, static on the communications links), but their cumulative effect fills Vashti with anger. The anger has no outlet because she feels helpless to do anything about these consumer complaints. She is even more helpless when the whole system begins to collapse.

If warnings about the degenerative effects of technological environments now seem misguided, it is largely because their target was never really a technical one; it was a political one. The real culprit in many of these analyses was not so much the technological paradise as the socialist paradise. If socialism should triumph, it was feared, human beings would lose fundamental sources of courage and creativity. They would become uniform ciphers, since degeneration means simplification. When Bulwer-Lytton presents such arguments, his political bias is quite evident: ". . . where a society attains to a moral standard, in which there are no crimes and no sorrows from which tragedy can extract its aliment of pity and sorrow, no salient vices or follies on which comedy can lavish its mirthful satire, it has lost the chance of producing a Shakespeare, or a Molière, or a Mrs. Beecher Stowe."[69] Thus Bulwer-Lytton updates the paradox of Mandeville's "fable of the bees": without vice no civilization, without slavery no Harriet Beecher Stowe. Such paradoxes have always had more appeal to book-reading aristocrats than to slaves. Wells, Forster, Tarde, and Orwell are not so obviously retrogressive. The particular strength of their fiction comes from their ability to question goals with which they are fundamentally in sympathy.[70] Tarde chides socialists for having the right goals but choosing the wrong means. The passage from *Wigan Pier* cited just above appears in a long and thoughtful discussion in which Orwell urges socialists not to let their cause be equated too closely with mechanical progress, but rather to let it be associated with the progress of social justice and liberty: "I want a civilization in which 'progress' is not definable as making the world safe for little fat men."[71] According to Orwell, unless socialists adopt more challenging ideals, they will not be able to combat fascism, which incorporates, along with much evil, an admirable emphasis on discipline and toughness and a revolt "against hedonism and a cheap conception of 'progress.'"[72]

In the 1930s, Orwell's major concern was to make socialism less vulnerable to fascism. His worry that socialism might be interpreted as enervating, however, did not disappear along with the immediate fascist threat. This worry, which was repeatedly expressed by nineteenth-century proponents of capitalism, persists to this day. The complaint goes like this: Competition is the key to social progress, but socialism discourages individual initiative. Socialism coddles people by offering them too much social security from the cradle to the grave, and by accustoming them to welfare and other forms of state aid. The capitalist system is preferable because it toughens individuals rather than weakening them.

Criticizing a physical environment for offering too much security distracts attention from the real issues. Both technological and natural environments offer threats and challenges. Danger and security, however, are as much social categories as environmental ones. In practice, capitalism has been far more bent than socialism on providing technological comforts, at least in the form of marketable goods. The social hazards are what socialism, more than capitalism, seeks to eliminate through programs to deal with unemployment, medical expenses, old-age care, child care, and education.

Can social perils be eliminated without an oppressive degree of social organization and control? When the question is posed in this way, we approach the real objection to the artificial paradise: not that it would be too comfortable, but that it would be too oppressive.[73] The enclosed, claustrophobic underworld serves as an objective correlative not so much for the triumph of technology as for what Irving Howe has called "the triumph of the Collective."[74] People are tamed along with nature. All these tales exemplify Lewis Mumford's dictum that "to the extent that men have escaped the control of nature they must submit to the control of society."[75]

In these frighteningly cohesive societies, both internal and external foes are subjected to this control. The vril-ya are willing and able to destroy any threatening outsider with a bolt of vril. (The narrator trembles, "I could not banish from my mind the consciousness that I was among a people who, however kind and courteous, could destroy me at any moment without scruple or compunction."[76]) In "The Machine Stops," when Kuno reaches the earth's surface he sees the bones of those who were condemned to perish there after the Great Rebellion. In *Underground Man,* a series of terrible wars eliminates first the industrial and then the free-thinking cities, and the threatening Chinese are walled out forever. All these are sanitized versions of Hay's genocidal wars. They all rest on the same assumption: that the establishment of a stable, harmonious society first requires getting rid of those who are different or defiant.

Once established, the society preserves its internal stability through a highly authoritarian social system. There are no recognizable governments in these tales. These societies represent the end of ideology, the end of politics, the end of change and conflict. They are more like churches, grounded on common beliefs and governed by priests, than they are like states, based on laws and governed by politicians. In "The Machine Stops," social control actually evolves into a Religion of the Machine, complete with a sacred Book of the Machine and various rituals of praise and prayer. Vashti is so imbued with this faith that she assumes the Central Committee of the Machine must be right, even if it means her own son's death.

Similarly, in the "benevolent autocracy" or "aristocratic republic"[77] of the vril-ya, the only formal authority is the Tur, who is obeyed without question (a three-person College of Sages assists him on the rare occasions when he is perplexed). No one considers the Tur's decrees unwise or unjust—"We do not allow ourselves to think so, and indeed, everything goes on as if each and all governed themselves according to immemorial custom."[78] The primary form of government is thought control. The expression for something illegal or forbidden is simply "It is requested not to do so-and-so." The narrator explains: ". . . though there were no laws such as we call laws, no race above ground is so law-observing. Obedience to the rule adopted by the community has become as much an instinct as if it were implanted by nature. . . . They have a proverb, the pithiness of which is much lost in this paraphrase, 'No happiness without order, no order without authority, no authority without unity.'"[79] In Tarde's *Underground Man,* once again, social unity is based on a shared faith—"the true religion" of love—and the rulers, again, are a natural autocracy. In the "Geniocratic republic" of the neo-Troglodytes, there is always a "superior genius who is hailed as such by the almost unanimous acclamation of his pupils at first, and next of his comrades," and no one feels resentful when such a genius becomes the supreme magistrate.[80]

A prime reason why all these societies need strong social authority is their need to limit reproduction. The underground setting is a test case for population control, because it provides a model of an environment that is strictly limited in terms of resources and waste-disposal sites. A strict population policy is therefore necessary to ensure the stability of a subterranean society. Tarde's neo-Troglodyte narrator asks in mock surprise:

Is it possible that after manufacturing the rubbish heaps of law with which our libraries are lumbered up, they precisely omitted to regulate the only matter considered worthy to-day of regulation? Can we conceive that it could ever have been permissible to the first comer without due authorisation to expose society to the arrival of a new hungry and wailing member—above all at a time when it was not possible to kill a partridge without a game licence, or to import a sack of corn without paying duty?[81]

The neo-Troglodytes require celibacy except under strictly defined conditions, the vril-ya forbid marriage with a person of another race, and the Central Committee enforces eugenicist policies that put to death any child showing undue strength and that forbid Kuno to be a father. In all cases, the regulations are enforced primarily by the force of opinion. Disobedience is almost unimaginable. For the few who can imagine it, the punishment is exile, homelessness, banishment.

Only the degenerate can tolerate such oppression. The nemesis of the artificial paradise is the loss of autonomy. Where resources are strictly limited, so must individual passion be. Where a high degree of organization is needed, obedience and practicality are more functional than energy, imagination, and initiative. The capacity to think and act independently is unwelcome. Bulwer-Lytton tells us of the vril-ya that "if you would take a thousand of the best and most philosophical beings you could find in London, Paris, Berlin, New York, or even Boston, and place them as citizens in this beatified community, my belief is, that in less than a year they would either die of *ennui,* or attempt some revolution by which they would militate against the good of the community, and be burnt into cinders at the request of the Tur."[82] The undegenerate are unregenerate opponents of the collectivity. They defy collective reason in the name of individual passion—either sexual passion, as in the case of Zee and the neo-Troglodyte lovers, or, as in the case of Kuno, a more general passion to use the body as an ultimate reference point of value ("Man the Measure," Kuno keeps telling himself). The only form their resistance takes is flight from the collectivity. In the underworld, flight usually means a journey to the surface. So Zee asks the narrator of *The Coming Race* to escape with her, either to another part of the underworld or to his surface world; so Kuno clambers up through the dark tunnels and shafts to emerge into the low hills of Wessex; and so the neo-Troglodyte lovers emerge on the summits of volcanoes to die in pairs.

Because the journey to the surface is a retreat from civilization toward nature, it is similar to the journey from the city to the woods that is a fundamental motif in American literature. (Indeed, when Zee beseeches her human lover to escape with her, her words echo those of Hester Prynne begging the Reverend Dimmesdale to fly with her into the wilderness.[83]) The prime reason for the retreat is not to trade one physical environment for another, but to find relief from an oppressive social environment. According to Leo Marx, "In imaginative literature . . . the concept of 'the city' must be understood as in large measure an abstract receptacle for displaced feelings about other things. . . . Anti-urbanism is better understood as an expression of something else: a far more inclusive, if indirect and often equivocal, attitude toward the transformation of society and of culture of which the emerging industrial city is but one manifestation."[84] The basic movement of these stories, then, is not so much away from technology and toward nature as it is away from social oppression and toward individual fulfillment.

But we should not be too reductionist here. If the contrast between the underworld and the surface is primarily a contrast between collectivity and liberty, this does not mean that the contrast between the artificial and the natural environment is nothing but a social metaphor. Forster in particular would object to such a reading of "The Machine Stops." When Kuno emerges onto the surface and sees a natural landscape for the first time, he lies on the grass and feels "the sun shining into it, not brilliantly but through marbled clouds,—the peace, the nonchalance, the sense of space."[85] Forster stresses the difference between this experience and the "artificial air, artificial light, artificial peace"[86] to which Kuno awakens after he has been dragged back down to his room. The term *artificial* is applied seamlessly to the external environment (air and light) and to the inner world of feeling (peace). How is genuine peace, as opposed to the fake kind, linked to the sun and clouds? Kuno is aware of the historical associations of the hills of Wessex, for he has heard a lecture on the kings who once ruled there. He also knows that other mountain landscapes are far more dramatic. "But to me," he tells his mother, "[the hills of Wessex] were living and the turf that covered them was a skin, under which their muscles rippled, and I felt that those hills had called with incalculable force to men in the past, and that men had loved them."[87] His response is not so much aesthetic, at least in any narrow sense of the word, as it is moral and ultimately political. Kuno admires not precisely the beauty of the landscape, but its power: the power of life (as opposed to mechanism), the power of the past, his own power as an individual. For him Wessex is not scenery but a muscled, communicative force. Kuno seems to end his description of Wessex with a purely "aesthetic" response: "I forgot to mention that a belt of mist lay between my hill and other hills, and that it was the colour of pearl."[88] Toward the end of the story, however, we discover that even the fragile beauty of the mist is significant to Kuno as a source of power. This cloud, we learn, shelters the Homeless who dwell on the surface, the saving remnant of humanity. So he and Vashti die, but by recalling the landscape of Wessex, Kuno is convinced that they have "recaptured life."[89]

So nature, to Forster, is not just a metaphor for individual freedom. Nor is nature a gym for working out to prevent physical deterioration; Kuno develops his strength by walking and doing calisthenics while he is still in the underworld. Nature's significance is, rather, as an independent and indispensable moral reference point. It provides a sense of connection between human and nonhuman life, between present and past and future. In the subterranean technological world, the only available measures of values are the artificial and the present. To Forster these standards, as opposed to "man the measure," are inherently oppressive. What is unique about the natural environment, what can never be replaced by the technological one, is its independence of the social order.

From this viewpoint, the concept of environmental politics takes on new meaning. The very existence of nature, its sheer being, challenges the life-denying, past-denying values of a machine-oriented collectivity. Because nature itself escapes from social determinism, it offers a reference point so human beings too can escape from social determinism. Forster writes in a pastoral tradition that "asserts or implies a continuity between the human spirit and the natural universe that is distinct from social definitions or placements of character."[90] Were nature hidden or destroyed, people would have no independent source of value by which to judge the dominant order. To appeal to the physical world is to appeal to a moral and political counterforce.

In that case, the real danger to the environment is not desacralization but destruction. To Forster, nature cannot be desacralized. That is precisely why the Committee of the Machine wants it destroyed—to get rid of an opposing source of authority. Pollution, then, is to be taken seriously in its ritual meaning as desecration. The disappearance of the untainted, the untouched, is not just an example of lamentable but correctable mismanagement. The destruction of nature is also an expression of powerful, possibly ineradicable aggressive impulses. "If man *must* conquer nature in order to survive," Daniel Callahan writes, "he also *wants* to conquer nature as an expression of his omnipotence. . . ."[91] Natural despoilation is not just a result of economic pressures; it is also a political action aimed at removing a source of subversion.

Forster gives some examples of the Machine's not-at-all-wanton destruction of nature. When Vashti flies in an airship above the Himalayas, she reluctantly glances out the window and sees where "the forests had been destroyed during the literature epoch for the purpose of making newspaper-pulp."[92] Another passenger mentions how in ancient times those peaks were considered "an impenetrable wall that touched the stars,"[93] a place where only gods could exist. The contrast is painful. Similarly, the Mending Apparatus denudes the entire Wessex hollow in which Kuno has taken refuge, ripping up all the brushwood and ferns until it maims him too. Its destruction of nature conveys the message that the power of the Machine is unassailable. Nature is destined to become a ruin, a part of ancient history, a dead empire of which we now see the half-buried relics. Like decaying bodies on the gibbet, rotting rivers and ruined meadows are cautionary lessons, the emblems of superior power. Nothing is sacred but that power.

Therefore, mourning for the forgotten or ruined landscape is not necessarily a sentimental and outdated emotion. It can also express an ideological response to tyranny. At the time Forster was writing, to be sure, there was considerable mourning for the landscape that could fairly be described as anachronistic. Not so much in Britain (where the sturdy yeoman was already a dim and distant memory), but more on the Continent (where a large percentage of the population still lived in rural areas), laments over the loss of human ties to the land mounted in the 1890s and the early 1900s. In France, the "integral nationalist" Maurice Barrès wrote *Les Déracinés* [The Uprooted] (1897), an immensely influential novel decrying the decadence that supposedly results when the peasant leaves the land. In Germany too there was much "cloudy moralizing about the *Volksgeist*"[94] and about the supposedly fundamental link between blood and soil. The influence of such ideas is evident in *The Decline of the West,* where Spengler laments the loss of the "profound affinity" between human beings and earth in the age of civilization and giant cities. Civilized man, like primitive man, is a nomad, "wholly homeless, as free *intellectually* as hunter and herdsman were free sensually"; "the rootless intellect ranges over all landscapes and all possibilities of thought."[95]

Like many reactionary appeals to the soil, "The Machine Stops" culminates in images of blood, death, and graves. Yet, as we have also seen, the story is remarkably prophetic in forecasting the ideological implications of environmental pollution. Like Villiers, Forster is a man of yesterday who also looks forward to tomorrow. To borrow terms introduced by Raymond Williams, concern about the loss of connection between human beings and the landscape can take either a residual or an emergent form, and sometimes a mixture of the two. A residual form represents elements of past culture that are largely excluded in the dominant culture. In this case it is a major element: the rapport between human beings and the nonhuman environment that arises from their daily interaction. Emergent forms, on the other hand, represent "new meanings and values, new practices, new significances and experiences . . . continually being created."[96]

A residual form of environmental politics seeks to keep the peasants on the land, in order to create a docile, conservative bloc of voters amenable to the leadership of the landed gentry. An emergent form of environmental politics—and this is what Forster's story primarily suggests—recognizes natural destruction as a requirement, rather than an accidental byproduct, of a social and political system that excludes from its moral frame of reference values that are not material, utilitarian, and presentist. In this context, the defense of nonhuman nature is part of a larger insistence upon the creation of a larger, richer frame of human values. Nature provides not a means of escape but a means of articulation. In *For a New Novel* (1963), Alain Robbe-Grillet summarizes the essence of Forster's belief in these words: "Belief in a *nature* thus reveals itself as the source of all humanism, in the habitual sense of the word. And it is not accident if Nature precisely—mineral, animal, vegetable Nature—is first of all clogged with an anthropomorphic vocabulary. This Nature— mountain, sea, forest, desert, valley—is simultaneously our model and our heart."[97]

If, as Robbe-Grillet suggests, antiquity is characterized by a sharp distinction between the natural and the human realms, the modern age is characterized by the interpenetration of natural and social categories.[98] The idea that nature has inherent moral value emerged during the eighteenth century, under historical conditions dominated by the development of industrial capitalism. This idea was expressed, among other ways, in the aesthetic of sublimity, which invested with moral significance the experience of an individual who confronts the superior power of nature. This emphasis on power invested nature with social significance. As the concept of sublimity was gradually transferred from nature to technology, human-made artifacts tended to be regarded as natural elements—once again an interpenetration of natural and social categories. In this chapter we have seen that society itself can be viewed as a force of nature, as a sublime power. By the later 1800s the social landscape, even more than the natural one, aroused a sense of fatal helplessness before superior power. Society itself was viewed as an "artificial infinity," an endless, mind-numbing spectacle that overwhelmed the individual with a sense of his comparative unimportance and impotence.

We have entered the age of the social sublime. In *The Coming Race* the terrible force of the vril-ya's deadly virtue arouses in the narrator "a profound terror" and "a moral awe" before these "mysterious powers."[99] The neo-Troglodytes bow down before a collective god, the religion of love. For Vashti in "The Machine Stops," the Machine is the power that commands veneration. Even before the establishment of the Religion of the Machine, she worshiped its miraculous powers and kissed the Book of the Machine in a "delirium of acquiescence."[100]

What is so interesting about Vashti, though, is that she, as much as Kuno, feels the need to retreat from the overwhelming power of the Machine. For her, society is a savage wilderness. She is terrified of the public sphere; she knows several thousand people, but she trembles to venture into a crowd. The thought of visiting her son makes her exceedingly nervous and frightened. When she finally ventures onto the airship, she is irritated by having to submit to the glances of other passengers, by the supposed unfairness of the flight attendant, by the cry of another woman whom she touched accidentally, and by being so close to other bodies. "People at any time repelled her. . . ."[101] Her isolation, as much as her physical weakness, makes her politically helpless. When the Machine begins to break down she becomes irritated, but it never occurs to her to consider organizing collective action. She files a complaint and retreats into her media-filled cell. In that enclosed room, at least, she has the sense of being in control. There, gadgets soothe her agoraphobia.

The appeal of consumer culture, the lure of the enclosed artificial paradise, is in part a response to the sad spectacle of ruined nature. Another part of its appeal, though, is that it serves as a refuge from the ruins of the public realm. The artificial paradise of the consumer, the domain of technological magic, is a charm against the artificial infinity of the social sublime. The highly developed consumer environment is at once a response to and a cause of the degeneration of social life.

Journeys into the Social Underworld

6

But what about the Morlocks? As we have just seen, a dominant anxiety in the late nineteenth century was that the individual would be crushed by the collectivity. Though commonly expressed in the biological language of degeneration, this anxiety was primarily a political one. Another fear—also expressed in biological terms, but also essentially political—was that society would become even more radically divided between rich and poor. In that case, not just a few dissidents but an entire class might become restive. Defiance might not end with personal rebellion, but might lead to general revolution.

When *The Time Machine* was published in 1895, the class split it portrays was the central social issue. It was by no means a new issue, of course. A fundamentally vertical social topography, with a small group of the rich on top and a large group of the poor beneath, had been a fact of social life as far back as historical memory ran. Until the nineteenth century, however, the poor were not considered a cohesive group—an "interest" (in England) or an "estate" (in France). Even though the term "the lower class" was used before 1800 in England, it was an umbrella expression covering a wide variety of social types: the laboring classes, divided from one another by geography and occupation (most notably the division between rural peasant and urban worker), and the "dangerous" or criminal classes, isolated and incohesive by definition.

The industrial revolution did not create the division between rich and poor, but it recast it as the division between capital and labor. In England the term "working class" emerged only around

1815, and by the 1820s "class" was a familiar social label.[1] It was only in the 1840s, however, that the concept of a specifically industrial working class took shape under the pressures of Chartism. In that decade the concept of "two Englands"—the privileged and the deprived, now identified with those who owned and those who worked—became part of the national vocabulary.

In France, where industrialization came later and less intensely, awareness of a lower class specifically associated with factory labor did not fully emerge until the Third Republic. In that nation, the great source of social upheaval was not the industrial revolution but the French Revolution. Indeed, the metaphor of an "industrial revolution" was invented in France early in the nineteenth century to express the affinity between the unprecedented rate of technological change and the great political revolution of modern times. Together the two new forces, political and technological, threatened the old order everywhere. Both represented a new "claim of the propertyless, working masses for a fair share of the necessities and perhaps even the felicities of life."[2] While the industrial revolution highlighted the vertical divisions of society, it also offered for the first time the promise of ending such stratification through greatly increased productivity. The French Revolution proclaimed the goal of democracy; industrialization promised to provide the means.

From the point of view of established society, the two revolutions represented a dangerous eruption from the social depths. A new type of poor, the industrial working class, spilled onto the social landscape. Because factories were isolated from the rest of society, and because so many workers were concentrated there, the proletariat had a mass identity that "the poor" or "the lower classes" had never had—an identity that inspired in Karl Marx the conviction that this class would develop the political consciousness and the organizational discipline to bring about a socialist revolution. Many people who did not share Marx's enthusiasm for such a revolution nonetheless shared his assumption that the factory was a political environment as well as a technological one. In that environment, many feared, workers isolated from the rest of society might be degraded into a hungry mob bent on devouring those better off: Morlocks.

But such fears were mingled with sympathy. If the factory promised to strengthen the collective power of workers, as individuals it weakened them. Ford Madox Brown's painting of heroic navvies digging up a London street should be compared with contemporary photographs of machine-tenders in the mills of Manchester. The navvies vigorously swing picks and shovels in the open air, surrounded by middle-class onlookers; the factory workers stand passively inside massive enclosures, shut off from nature and society alike. Instead of being strong-muscled, "newfangled men,"[3] factory workers were often smaller, frailer women and children. In the factory environment, the mindless and the weak seemed favored to survive.

In that case, who are the Eloi and who are the Morlocks? Not the leisured classes but the laboring ones suffered the most physical degeneration from the advent of industrialization.[4] There are few historical documents more heartbreaking to read than the records of the government commissions in mid-nineteenth-century England that inquired into the physical effects of industrial labor. Parents tell of watching helplessly as their children became progressively more twisted and deformed from their labor. The adults themselves felt their lungs being devoured by the dust-laden air, and old and young alike worked in constant fear of accidents that could maim or kill.[5]

Considered as a place of production, then, the technological environment had contradictory effects. It turned workers into both Eloi and Morlocks. It wore them down, weakening and even killing them; at the same time, it promised to empower them. Helpless suffering and revolutionary upheaval were the conflicting possibilities presented by the new industrial workplaces. Sympathy and fear were the conflicting responses of those on the higher levels of society.

One of the great projects of secular literature, as it has developed in the West since the eighteenth century, has been to reconcile sympathy and fear through narrations in which the reader is escorted into the lower depths of society.[6] The common structure of these narrations is a movement from the comfortable but deceiving social surface down into the grim, dark world of the underclass. The very shape and movement of the narrative is political, "because it evokes both social wishes and social fears and then negotiates among them, establishing fictional paths through highly charged ideological territories."[7]

The development of literary realism is usually interpreted as part of the great transformation from sacred to secular, from myth to reason, that reshaped Western culture from the seventeenth century on. Yet once again a remnant of the mythological lurks in the halls of modernity. The nineteenth-century tradition of literary realism incorporates a powerful mythological tradition: the story of a journey to the underworld in quest of truth. In realistic literature, the pilgrim descends into the social depths in search of social truth. The descent is always metaphorical, but in view of the living conditions of the poor it may be literal as well.

Consider, for example, this subterranean journey described by Daniel Defoe, one of the great originators of realism, in *A Tour thro' the Whole Island of Great Britain* (1725). He tells how he and his traveling companions decided to seek out a natural wonder of the Peak District called the Giant's Tomb, a high mountain where a giant was supposedly buried. The party clambered up the mountain in search of this spectacle of natural sublimity, but "miss'ed the imaginary Wonder, and found a real one." Defoe then tells in advance what will be the moral purpose of this story: "to shew the discontented part of the rich World how to value their own Happiness, by looking below them, and seeing how others live." The travelers discover a small opening in the mountain rock that, to their amazement, turned out to be the home of a family with five children. When they inquire if they may enter, the mother invites them in although "'tis not a Place fit for such as you are to come into." They find themselves in a "large hollow Cave" divided into rooms by curtains. Defoe reports that "Every Thing was clean and neat, tho' mean and ordinary." The father, they are told, works in a nearby lead mine, as does the mother when she is not busy tending the children. Together they earn at best eight pence a day, and yet they manage to live with dignity. The spectacle so moves the visitors that after a short conference they give the lady "a little Lump of Money." She bursts into tears, thanks them profusely, and tells them how contented she and her husband are despite their poverty. Defoe again stresses the moral import of his narration, and again invites his readers to share the moral education he has gained: "In a word, it was a Lecture to us all, and that such, I assure you, as made the whole Company very grave all the rest of the Day: And if it has no effect of that kind upon the Reader, the Defect must be in my telling the Story in a less moving manner than the poor Woman told it her Self."

The travelers then follow the woman's directions to the nearby lead mine. As they stand looking at the narrow mouth of one pit, they are "agreeably surprized with seeing a Hand, and then an Arm, and quickly after a Head, thrust up out of the very Groove we were looking at." After this "subterranean Creature" extracts himself from the pit, they ask him about his work, the depth of the mine, and his tools. Defoe concludes by addressing the reader one more time:

If any Reader thinks this, and the past relation of the Woman and the Cave, too low and trifling for this Work, they must be told, that I think quite otherwise. . . .

. . . If we blessed our selves before, when we saw how the poor Woman and her five Children lived in the Hole or Cave in the Mountain . . . we had to acknowledge to our Maker, that we were not appointed to get our Bread thus, one hundred and fifty Yards under Ground. . . . Nor was it possible to see these miserable People without such Reflections, unless you will supposed Man as stupid and senseless as the Horse he rides on. But to leave Moralizing to the Reader, I proceed.[8]

But of course Defoe is not leaving the moralizing to the reader. Defoe and his successors act as a tour guide for the reader who is unacquainted with the social depths. What the reader sees down there (the writer-guide assumes) will inevitably lead to a spiritual awakening: gratitude for one's own lot, respect and charity for those less fortunate. The narration is also a "Lecture."

Writers in the realist tradition apply to literature the Baconian assumption that "truth lies hidden deep in mines." The social investigator, as much as the natural one, must dig downward to find the truth—in this case, the truth about the poor. They are intellectually problematic, the subject of investigation by excavator-researchers. In England the great mid-Victorian realists, such as Dickens, Thackeray, and George Eliot, were routinely praised for their detailed and highly accurate descriptions of social life; the words *copy, transcript, photograph,* and *daguerreotype* were used by both defenders and critics of literary realism. Factual accuracy was not an end in itself, however. The ultimate end of the realistic description was moral education.

The outstanding example of such a moralist-realist is of course Charles Dickens. As one critic has remarked, Dickens writes like an excavator who makes a vertical cut into society.[9] Dickens's primary concern is to show how all the strata are related. The fact that all of society is interconnected below surface appearances is for him the objective basis for an ethic of social responsibility.

Dickens, though, had the unique gift of discerning the fantastic in the seemingly realistic, and, conversely, of using fantasy to highlight reality. In *The Pickwick Papers* (1837), the book that launched his career, Dickens tells a story that is a surreal version of the social journey related so realistically by Defoe. The story tells how Gabriel Grub, a gravedigger and sexton, "an ill-conditioned, cross-grained, surly fellow—a morose and lonely man," goes to a churchyard on Christmas Eve (having first thrashed a little boy to stop him from singing Christmas carols). A group of goblins appear and pull the terrified Grub down into their cavern. There, beneath a tombstone, they in turn thrash Grub and show him two scenes to make him realize what a miserable creature he is. First, a cloud at the end of the cavern rolls away and reveals "a small and scantily furnished, but neat and clean, apartment" in which a mother and her happy children dwell. The father comes home from his work, wet and weary; as he sits down to his meal, the children climb on his knee and the mother sits by his side. The scene gradually darkens as the youngest child dies and then the parents, who have grown old and helpless. The remaining children, although mournful, retain their faith in the afterlife and their determination to find contentment and cheerfulness in the busy world.

The second scene reveals "a rich and beautiful landscape" of shining sun, sparkling waters, rustling wind, butterflies, and flowers. "Man walked forth, elated with the scene; and all was brightness and splendour." Thanks to these two spectacles, one of human and one of nonhuman beauty (and also thanks to sound thrashings by the goblins), Grub learns his lesson: "He saw that men who worked hard and earned their scanty bread with lives of labour were cheerful and happy; and that to the most ignorant, the sweet face of nature was a never-failing source of cheerfulness and joy. . . . Above all, he saw that men like himself, who snarled at the mirth and cheerfulness of others, were the foulest weeds on the fair surface of the earth; and setting all the good of the world against the evil, he came to the conclusion that it was a very decent and respectable sort of world after all." To see is to repent. Grub, now "an altered man," cannot bear to return to "a place where his repentance would be scoffed at and his reformation disbelieved," so he leaves to "seek his bread elsewhere."[10]

This tale is, of course, the prototype for Dickens's beloved *Christmas Carol* (1843), in which Grub becomes Scrooge and the goblins turn into three ghosts. In *A Christmas Carol* Dickens puts more emphasis on the sufferings of the poor, but the basic design is the same. The unhappy, selfish man who does not recognize his own blessings and who lacks sympathy with others is forced to journey below the surface of society. When he sees firsthand the sorrows and the courage of those who dwell in social darkness, his heart is transformed. As a writer, Dickens is the goblin or ghost. He takes the reader—who is not so bad as Grub or Scrooge, to be sure, but who is still in need of moral education—below the surface of society.

Two years after the publication of *A Christmas Carol*, Dickens's French contemporary Victor Hugo began work on a book called *Les Misères*, eventually published in 1862 as *Les Misérables*. Hugo, like Dickens, not only wants to show the reader the subterreanean life of society; he also wants to show the inextricable connections among the social levels. In Hugo's work, however, the writer leads the reader not into the subterranean world of the respectable poor, but into a much more dangerous and threatening underworld of misery. For Hugo, the tradition of literary realism means more than the moral education of the respectable. The poor whom he unearths arouse fright, not sympathy. For them, Hugo believes, the only solution is to change a social structure that condemns them to dwell in fetid darkness. He moves beyond moral lessons to political ones.

For Hugo the social underground, however evil and dangerous, is still part of the social whole. It therefore harbors the possibility of social change. Midway through *Les Misérables,* as the narration approaches the uprising of 1832, Hugo halts to meditate on the hidden depths of French society: "The social soil is mined everywhere, sometimes for good, sometimes for evil. These works happen in strata; there are upper and lower mines." He then describes the many varieties of undermining, which are hardly noticed by society and yet change its substance. "What comes from all this deep delving? The future." In the upper mines are the great political and revolutionary and philosophical excavations, which are heroic and which lead to progress. Deeper down, revolution becomes increasingly murky: ". . . below Condorcet is Robespierre; below Robespierre is

Marat; below Marat is Babeuf. . . ." Even farther down, disinterestedness disappears. "There begins evil. . . . There is a point where undermining becomes burial, and where light goes out." At the very bottom of society is "the ultimate hole." "An awe-inspiring place. . . . It gives onto the abyss." Here live savages, driven only by want, and criminals, the stepchildren of ignorance and misery. "This cave below everything is the enemy of all. It is universal hatred. . . . The object of this cave is the ruin of all things. . . . All the others, those above it, have only one object—to eliminate it. . . . Destroy the cave Ignorance, and you destroy the mole Crime. . . . The only social peril is darkness." Hugo then introduces four such "misshapen toadstools of civilization's substrata," four bandits known collectively as Patron-Minette. During the day, they sleep in plaster kilns or abandoned quarries or sewers; at night, they rob. When respectable people glimpse them at midnight, as dark forms on a lonely boulevard, "they are terrifying. They do not seem like men, but forms fashioned of the living mist: you would say that they are generally an integral portion of the darkness. . . ." This is not the poverty of passive suffering, not the devoted wife and children and hard-working man who are the deserving poor. These are the dangerous poor, the anaerobic sludge of civilization. It is no remedy simply to show this poverty to middle-class people. Instead of being moved, they will only be frightened and perhaps mugged. "What is required to exorcise these goblins?" Hugo asks, and responds to his own question: "Light. Floods of light. No bat can resist the dawn. Throw light on the society below."[11]

  Les Misérables is saturated with memories of the French Revolution. When Hugo asserts that "the revolutionary sense is a moral sense," he expresses a moral sensibility very different from that of Defoe or Dickens. In Britain, Defoe is startled to see a single lead miner suddenly appear from a hole in the ground. The French remember a far greater collective event, when (in Hugo's words) "men heard beneath their feet the obscure course of a muffled sound, when some mysterious uprising of molehills appeared on the surface of civilization, when the earth fissured, the mouths of caverns opened, and men saw monstrous heads spring suddenly from the earth."[12]

  It is never far from Hugo's mind that such an uprising might happen again. As Marius makes his way toward the barricades in the 1832 uprising, in a moment of calm before the revolutionary storm,

Hugo imaginatively soars far above Paris and sees the insurgent quarter of Les Halles as "an enormous black hole dug out of the center of Paris . . . an abyss." Only a few ghostly shapes glimmer in that "monstrous cavern . . . a savage darkness, full of snares." Yet in this void "the gloomy voice of the people was heard dimly growling. A fearful, sacred voice, composed of the roaring brute and the speech of God, . . . which comes at the same time below like the voice of the lion and from above like the voice of thunder."[13]

The climactic "chase scene" of *Les Misérables,* in which Jean Valjean carries the wounded Marius through the sewers after the fall of the barricade, is deservedly a famous moment in literature. An exciting chase in its own terms, it also reverberates with associations both Biblical and classical: Jean Valjean's descent into Hell, his need to bear his own cross, his entrapment in the mire, his glimpsing the light, his despair when he sees his exit blocked, his final redemption and ascent back to the land of the living. The sewer system of Paris becomes the setting of a timeless journey of initiation and expiation. Again, though, the journey differs in moral import from those described by Defoe and Dickens. Jean Valjean is no middle-class visitor being introduced to the social depths. An ex-convict, he has lived in that underworld himself. He has been an outcast, and he has made several earlier journeys to enclosed places (the prison, the cloister) in his never-ending flight from injustice. In his dual identity as Jean Valjean and Monsieur Madeleine, he incarnates the hidden but inseparable connection between the respectable surface world and the fearful subterranean world of criminality.

As the nineteenth century progressed, the conviction that such a connection existed began to disappear. The sense of disjuncture first becomes evident in the famous cluster of "industrial novels" published mainly in the 1840s in Britain.[14] To be sure, the writers of these novels were still intent on showing the suffering of the poor to the reader. In *Mary Barton,* in the chapter titled "Poverty and Death," John Barton and his friend discover a starving family in a cellar. In *Alton Locke,* Sandy Mackaye guides the idealistic poet Locke on a sort of sightseeing tour of working-class London ("A ghastly, deafening, sickening sight it was. Go, scented Belgravian! and see what London is!"[15]) But whereas the urban tradition in realistic fiction favored a sweeping panorama, rich in details and variety, so that the city was presented a complex but organic whole, the industrial novel encouraged the establishment of a class image. Variety and quirkiness were replaced by class solidarity and solemnity; the complexity of London yielded to the blank uniformity of Northern cities.[16] For example, the streets of Coketown in Dickens's *Hard Times,* unlike the wildly varied byways of London, are filled with faceless, homogeneous workers.

This more uniform view of the social depths encourages fear and withdrawal rather than sympathy and charity.[17] The poor are picturesque, but the proletariat is sublime. When factories and mines are described as landscapes from hell, factory workers and miners are then readily seen as demons or goblins—as frightening, subhuman, subterranean creatures. And when there are so many of them, when the rows of workers' houses seem to stretch on forever, when a sea of faces erupts from the factory at the end of the shift, the infinite vista that stirs the sublime response of fear and awe is now

"Capital and Labour," a cartoon from *Punch* (volume 5, 1843). The caption states, with heavy sarcasm, that the cartoon is intended to do away with "very offensive representations of certain underground operations, carried on by an inferior race of human beings [miners] . . . by showing the very refined and elegant result that happily arises" from their labor. "The works being performed wholly underground, ought never to have been intruded upon the notice of the public." Courtesy of Lamont Library, Harvard University.

that mob. Such images are used repeatedly in the industrial novels of the 1840s, most particularly in scenes where the diabolical and infinite merge, where the mob is seen by the red glare of torches. The social abyss becomes, in Hugo's words, an "awe-inspiring place"; there the observer feels overwhelmed by a power far greater than his own.

In the Britain of the 1880s and the 1890s, East London was often described in this language of social sublimity. It seemed an immense city in itself, set apart from the rest of London, an endless vista of poverty, with "sprawling, seemingly never-ending streets of poor houses and people."[18] This topographical separation reflected a more general sense that the working poor were an isolated, outcast group. The East Londoners were compared to Polynesian or African tribes, or (by Jane Addams) to the salt miners of Austria. Some still made the journey to these social depths. In his study of the working classes in Victorian fiction, P. J. Keating calls settlers, salvationists, and missionaries "new link heroes" and adds that they "figure prominently in the slum fiction of the nineties."[19] The point is that now they are heroes. The ordinary soul, however well-intentioned, would be reluctant to enter the self-enclosed environment of the down and out.

If Dickens makes vertical cuts through society, later nineteenth-century novelists (e.g. George Gissing and Émile Zola) make horizontal ones. In their novels the poor are cut off from the rest of the world. Trapped in the environment of poverty, unable to work their way up to the social surface, unable to breathe healthy air, the poor degenerate. Gissing and Zola stress the pathologies bred in the social underworld: disease, sexual license, alcoholism, violence. As these pathologies become inbred, the poor seem to become another species entirely. The classic study of such degeneration is Zola's *L'Assommoir* (1877, published in English in 1883), which traces the moral and physical decline of the Coupeau family. Zola was greatly influenced by the French tradition of medical degeneracy; in *Le Roman Expérimental* (1880), he even compared his aims as a novelist to those of the physician Claude Bernard. Gissing, on the other hand, was less interested in heredity and instead emphasized the slum environment as a causal factor. Both Gissing and Zola, though, assume that inexorable forces trap the poor despite any flutterings of virtue or conscience.

At that point literary realism mutated into naturalism and became identified with descriptions of degeneration: alcoholism, family breakdown, disease, crime, death. For naturalists, real life is low life. In that case they had little reason to believe, as realists always had, in the implicit connection between showing the truth and imparting moral education. When the social depths are so dark and vast, they seem not a human environment at all but a wilderness or jungle (to use metaphors commonly applied to "darkest London"). The novelist may still guide the reader downward through the social strata, but what they see there is so monstrous that the reader's response is more likely to be revulsion than sympathy.

In Gissing's *The Unclassed* (1884) the characters themselves allude to this problem when they discuss the art of fiction. One of them, the writer Osmond Waymark, expresses Gissing's views: "The fact is, the novel of every-day life is getting worn out. We must dig deeper, get to untouched social strata. . . . [My book will be] for men and women who like to look beneath the surface, and who understand that only as artistic material has human life any significance." Another character later praises Waymark's novel in terms that gratify the author: "'It is horrible,' he exclaimed; 'often hideous and revolting to me; but I feel its absolute truth.'" The reader then adds that "such a book will do more good than half a dozen religious societies."[20] But how can the reader who is shocked and sickened conclude that the poor deserve charity? To be sure, if the working-class environment is so radically different from his own, then applying his middle-class values to it is futile. But what other values is the middle-class reader going to have? His desire to withdraw only grows stronger.

This seems to have been the response of Gissing himself. In some earlier works [*The Unclassed, Workers in the Dawn* (1880), *Demos* (1886)] he pinned his hopes on working-class intellectuals who use cultural improvement as a means of escape from the slum. In *The Nether World* (1889), everyone is trapped. Its inhabitants can only try to survive down below, at the best retaining some modicum of dignity amid the crushing forces. Gissing even gives up on descriptions of the slum environment: "Needless to burden description with further detail; the slum was like any other slum; filth, rottenness, evil odours, possessed these dens of superfluous mankind and made them gruesome to the peering imagination."[21] *The Nether World* was

Gissing's last attempt at working-class fiction, except for a few short stories written at the request of an editor. Instead, he began to write about classless intellectuals who have no connection with working-class life. These heroes too find themselves overwhelmed by hostile social forces, but at least they are not vile. As Gissing explained in a letter, the only hope for such intellectuals to survive was to "make a world within the world."[22]

Here the realistic tradition of the subterranean social journey reaches a dead end. There is no point for the reader to read it, or for the writer to write it. To attempt a journey of connection is worse than futile. It might be positively harmful, inspiring in the middle-class reader an unjust but understandable revulsion against the miserable poor. The novelist, like his subjects, is trapped.

For the novelist, however, the trap is at least partly self-laid. For all their emphasis on truth-telling and realism, most novelists of working-class life in the 1880s and the 1890s were highly selective in their descriptions. Keating argues that the Victorian "realists" were decidedly unrealistic in choosing the poor and the debased as representatives of the working class. In *East London* (1889), the first volume of his *Life and Labour of the People,* Charles Booth estimated that only 1.2 percent of the population (which was just over 900,000) could be classified as "occasional labourers, loafers and semi-criminals." By far the largest group (42.3 percent) consisted of workers who received regular wages and lived above the poverty line. At least a third of the population, however, lived below that line. In short, two-thirds of the population, far from being debased, lived in relative respectability.[23]

Keating further argues that among English realists this evasion reflects a powerful but unacknowledged fear of class war. By concentrating on urban poverty, they either ignored or downplayed the potentially revolutionary events of those decades (serious strikes, an unprecedented growth of trade-union membership, the emergence of an independent socialist party). If the poor were not degraded in a hopeless mire of social pathology, they posed more of a challenge to the existing political order. The degeneration of urban outcasts may be repulsive, but it is relatively unthreatening. When the slum is a prison, then the rest of society is safe from its inmates (unless one is foolish or unlucky enough to venture into the wrong part of town).[24]

The key transition, as difficult to make now as then, is from social concern to political action. The realistic tradition in literature amply expressed social concern. That concern, however, tended to assume the continued existence of a vertically stratified social terrain. Hugo's revolutionary consciousness was the exception, not the rule. As the critic Michael Wilding reminds us, the emphasis on the realistic mode in surveys of the novel has tended to be a conservative stress. Realistic novels tend to present society both as a totality and as a finality.[25]

In *Politics and the Novel* (1957) Irving Howe makes a distinction between the "social" and the "political" novel. He proposes that in the tradition of literary realism, beginning with Jane Austen, there is a gradual shift from the former to the latter: ". . . the novelist's attention had necessarily to shift from the gradations within society to the fate of society itself. It is at this point, roughly speaking, that the kind of book I have called the political novel comes to be written—the kind in which the *idea* of society, as distinct from the mere unquestioned workings of society, has penetrated the consciousness of the characters in all of its profoundly problematic aspects. . . ."[26] But this emphasis on "the *idea* of society" would not develop exclusively, or even primarily, within the tradition of literary realism. Wilding emphasizes the importance of what he calls "political fictions" as an alternative mode that progressive writers have used to raise political and social issues.[27] Wilding goes on not so much to define political fiction as to list the various ingredients in what he considers an undefinable, mixed, but powerful mode: romance, fable, utopian fantasy, vernacular picaresque, and mystery, as well as realism (as that term is commonly understood). He argues that in the twentieth century the fictional examination of political issues has increasingly moved away from a naturalistic mode toward a fantastic and fabular one.

We shall now see how the journey to the social underworld can be presented in the latter mode. As politicized versions of the fictional subterranean journey, these are imaginary voyages into a strange but recognizable political terrain. This journey is not so much a thought experiment as what D. H. Lawrence called a "thought-adventure."[28] Its purpose is to discover not what is, but what might be. Two writers of this kind of thought-adventure—writers wildly popular in their own day and still widely read—are Jules Verne and H. G.

Wells. Today they are often considered forerunners of science fiction, but their more lasting claim to attention is that they were masters of political fiction.

Jules Verne's book *The Black Indies* (1877) opens with an invitation to a subterranean journey. Simon Ford, once a foreman at the Aberfoyle coal mine, sends a letter to James Starr requesting him to visit the pit the next day to receive "a communication of an interesting nature."[29] Starr, an engineer (and therefore a favorite type of Verne's, being "one of those practical men to whom is due the prosperity of England"), had managed the Aberfoyle mine for twenty years. Ten years earlier the last ton of coal had been extracted from this colliery, leaving an exhausted mine "like the body of a huge fantastically-shaped mastodon, from which all the organs of life have been taken, and only the skeleton remains."[30] At the occasion of its closing, Starr had made a moving speech celebrating the unity that comes from common effort: "The work has been hard, but not without profit for you. Our great family must disperse. . . . But do not forget that we have lived together for a long time, and that it will be the duty of the miners of Aberfoyle to help each other. Your old masters will not forget you either. When men have worked together, they must never be strangers to each other again."[31]

Simon Ford, however, had decided not to leave the mine. Along with his wife and their son Harry he had built a cottage in the Dochart pit, 1,500 feet below the surface. There he had searched tirelessly for new veins of coal, for he was convinced that the wealth of the mine had not been exhausted. Verne calls Ford "the type of miner whose whole existence is indissolubly connected with that of his mine."[32] In his bond with the land, and also in his desire for independence (one of his reasons for living underground is to be out of the reach of rent and tax collectors), Simon Ford is a subterranean version of the prototypical French peasant.

Relying on peasant instinct, Ford has discovered indications of another coal seam just at the point where the rational engineer had ended his soundings. This is the reason for his message to Starr. When Starr arrives, he and Simon and Harry Ford blast open a wall in the mine and reveal an immense excavation, as large as Mammoth Cave in Kentucky, dotted with subterranean lakes and criss-crossed with veins of coal running like "black blood":

. . . the place could be inhabited by a whole population. And who knows but that in this steady temperature, in the depths of the mines of Aberfoyle, as well as in those of Newcastle, Alloa, or Cardiff—when their contents shall have been exhausted—who knows but that the poorer classes of Great Britain will some day find a refuge?[33]

The author's grandson, Jean Jules-Verne, has argued that this passage reveals the traces of Verne's original, more ambitious plan for the novel: to show "a whole population" dwelling underground. When Jules Verne outlined this theme in a letter to his publisher Jules Hetzel, however, Hetzel vetoed the idea, saying that the idea of an underground England was too farfetched.[34] So in *Black Indies* the Fords are joined in their underground life not by an entire nation but by a more modest contingent: James Starr, who is happy to return to his engineering work, and a number of former colliers, who are also happy to leave farming and to return to a mining job offering steady work at high wages.

Together, down in the pit, they build Coal Town, a collection of picturesque houses and a chapel at the edge of the subterranean Lake Malcolm. Tourists flock to see the brilliantly lighted underworld. As always, Verne stresses the fit between nature and technology, a relationship where the latter perfects the former ("The electric discs shed a brilliancy of light which the British sun, oftener obscured by fogs than it ought to be, might well envy").[35] The mine, far from being a forbiddingly inorganic environment, is a nurse, a mastodon, a body criss-crossed by black blood. Harry Ford explores its ponds in his canoe and even goes hunting there, birds having been introduced into the cavern to feed on the fish swarming in the dark waters. Verne even uses a version of a traditional pastoral image, the echo, to express the rapport between man and mine. Harry's friend Jack Ryan, who loves to sing, says that he prefers to work underground because the vaulted roofs "merrily echo" his song, whereas on the surface the sound just dissipates.[36] This mine is a pastoral landscape, a black India.

Another leading theme of Verne's—and it is inseparable from his belief in the harmony of nature and technology—is his social ideal of a community united in the direct exploitation of nature's inexhaustible riches. It is the bourgeois dream *par excellence:* human progress based on the mastery and possession of nature.[37] More specifically, it is the Saint-Simonian dream of fraternal cooperation in labor guided by scientific knowledge.[38] In New Aberfoyle the engineer, the foreman, the workers, and the generous company directors lie down together like Biblical lions and lambs. The subterranean world is set apart from surface society as a social ideal.

But Verne is much more than a bourgeois mouthpiece. In the words of Jean Chesneaux, one of the most perceptive commentators on Verne, he is a bourgeois in revolt. Verne's inner conflict and his turmoil are mirrored in the various subterranean levels of *Black Indies*. If Coal Town represents a social utopia lying below the "luxury and intemperance of the outer world,"[39] it also lies above a wilder, more threatening world farther down. There is another, even deeper level to the mine, and consequently to the story.

The existence of this lower level is revealed gradually, through mysterious occurrences and accidents. A second letter to Starr, arriving shortly after Simon's, tells him not to visit the mine after all. Rocks fall, or are perhaps thrown, during the search for a new seam. Most alarming, when Starr and the Fords are exploring the new vein, their lamp is broken and their exit route is sealed off. They are trapped for ten days before Jack Ryan happens to come looking for them, and they survive only because some mysterious person leaves food and water near them.

Harry Ford is convinced that the pit harbors both a friendly person and a criminal. "I mean to seek them both in the most distant recesses of the mine,"[40] he announces. Journeying even farther undergound, he discovers, at the bottom of a deep natural shaft, the child Nell. Only half-civilized, she is capable of speech but so frightened that she hardly communicates. Gradually she learns to trust humanity as she responds to the kindliness of the Coal Towners. Still, she never explains why she was in the mine or who else might be there.

Coal Town may be an ideal society, but it turns out to be a haunted one. An irrational specter begins to stalk the mine when Nell agrees to marry Harry Ford. At the moment she accepts his proposal, the mine is suddenly flooded by the loch lying directly above it; disaster is averted only narrowly. As the date for the wedding approaches, accidents happen more and more frequently. Finally, a threatening note signed "Silfax" provides a clue to the mystery. In the mines of old Aberfoyle, Silfax had been the last one to hold the job of fireman. In the days before the invention of the Davy safety lamp, the fireman had prowled ahead of the other miners, producing partial explosions with a blazing torch, so that firedamp was dispersed before it accumulated in quantities great enough to cause a fatal explosion. Evidently, says Simon Ford, "the constant danger of the business had unsettled [Silfax's] brain." Even before Silfax's mind was deranged, Ford further explains, he had "a fancy . . . that he had a right to the mine of Aberfoyle . . . so he became more and more savage in temper the deeper the Dochart pit—his pit—was worked out." He too must have discovered the new vein of coal. "With the egotism of madness," Starr surmises, "he believed himself the owner of a treasure he must conceal and defend."[41] Nell finally reveals that Silfax is her grandfather. He had been her only companion in the mine (except for his ferocious pet owl, who had taken a liking to her).

The threat of terrorism changes the social order in Coal Town. Measures are taken that "seriously interrupt the work of excavation"[42]—the productive work that has been the source of social unity. Coal Town becomes an armed camp. Sentinels are posted at the mine entrances; any stranger attempting to enter is taken to James Starr to account for himself. Every gallery is searched. Despite these measures, at the moment Nell and Harry Ford exchange vows, a rock ledge at the edge of Lake Malcolm gives way and plunges into the abyss, releasing jets of firedamp. On the lake, Silfax appears in a boat, standing upright, holding a lighted Davy safety lamp and shouting "The fire-damp is upon you! Woe—woe betide ye all!" As the firedamp spreads, Silfax smashes the glass of the lamp and hands the lighted wick to his owl, so the bird can fly high up into the dome and ignite the accumulating gas. But at the last moment Nell calls out to the owl, which obediently flutters to her feet. Foiled, Silfax throws himself into the lake and drowns. After a suitable period of mourning, the young couple lives happily ever after in Coal Town.[43]

In *Black Indies* degeneracy is found not among the coal miners, who act like proper bourgeois, but in a sub-subterranean madman. An insane possessiveness, a crazed insistence on the rights of private property, is exhibited not by the mine's owners but by Silfax. Thus, Verne criticizes the destructiveness of property values only indirectly. In the ideal pastoral landscape of Coal Town, the owners are invisible and beneficent, happily paying good wages. But down below, the greedy and irrational quest for property almost destroys the workers' paradise. In the socially conscious part of the story—the establishment of Coal Town—the issue of property is evaded. In the romantic subplot, property is central.

At first it seems that Coal Town is going to be a cohesive community, like those in the underworld fantasies of Tarde and Bulwer-Lytton. It is to be a society without politics, laws, property, or state—because all these means of resolving conflict are unnecessary in a community united by the common goal of conquering nature. The citizens are to live together in one big frictionless family, universally accepting the authority of wise elders like James Starr. Then this community is confronted by a madman, a "malignant and invisible being"[44] bent on its destruction. The complicated, dangerous environment of the mine proves highly vulnerable to terrorism. In order to keep the mine running and to prevent a disastrous accident, the community resorts to other forms of authority. Toward the end of *Black Indies,* we see the beginnings of a military state: armed guards, inspections, check-in procedures.

For Verne, reconciling technology with liberty is the central problem. As Chesneaux makes clear in his splendid analysis of Verne's ideology, the author of the "imaginary voyages" simultaneously held an assortment of political convictions—not only the Saint-Simonian dream of nature's conquest, but also a fierce nationalism, an equally fierce libertarian individualism, and a social conservatism. As Chesneaux says, it is impossible to describe Verne's politics in any systematic way, for these convictions are in many ways contradictory. Furthermore, his more radical ideas were largely hidden behind a conformist bourgeois lifestyle. Chesneaux presents Verne as a sort of closet anarchist; another critic, Pierre Louys, calls Verne an "underground revolutionary."[45] Verne is as difficult to classify as Villiers de l'Isle-Adam. They are both neo-mystics, loners, nonconformists, at once radical and conservative.[46]

Despite all his complexities, however, Verne keeps returning to two fundamental values: the subjugation of nature and the liberation of individuals. He does not see these as conflicting goals. His books typically feature the establishment of a "free environment" (a phrase much used by anarchists in his day) which is free precisely because it is highly technological. The recurring Vernian fantasy is that of a small group of free individuals who use science and technology to escape the conventional laws of property and of nations and to create an ideal community. For Verne, the creation of a technological environment allows a retreat from civilization and history.

The classic example, and one to which Verne himself attached special importance, is of course *Twenty Thousand Leagues under the Sea,* an earlier and far more impressive work than *Black Indies.* Captain Nemo is the individual who seeks liberty in the artificial environment of the submarine *Nautilus.* In *Twenty Thousand Leagues* we only get hints about Nemo's background, but in its sequel *The Mysterious Island* (1875) we finally hear his story. Nemo was Prince Dakkar, an Indian, the son of a Rajah, highly cultivated in European ways but burning with idealistic hopes for his own people. In 1856 he participated in the unsuccessful Indian Mutiny against the British. Disappointed in his hopes of shaking off British rule, sickened by the destruction of his family, and consumed with hatred for the so-called civilized world, Dakkar decided to use science instead of the sword. He built the *Nautilus* on a desert island in the Pacific and disappeared beneath the seas. Now Captain Nemo, he regards the world below the surface as the ultimate free environment. Here, and here alone, he can be truly free from despotism. Once again freedom is found in exile, but this time it is subterranean self-exile from surface tyranny: "The sea does not belong to despots. Upon its surface men can still exercise unjust laws, fight, tear one another to pieces, and be carried away with terrestrial horrors. But at thirty feet below its level, their reign ceases, their influence is quenched, and their power disappears. Ah! sir, live—live in the bosom of the waters! There only is independence! There I recognize no masters! There I am free!"[47]

One contradiction, of course, is that this lover of freedom is the absolute commander of the *Nautilus.* However, this kind of *ad hoc* leadership—based on Nemo's unquestioned expertise as "captain, builder, and engineer"[48] of the vessel—is quite permissible under anarchist principles, which exclude primarily the multipurpose, formalized, armed authority of the state. The real challenge to Nemo's authority comes from the three captives who are brought on board the *Nautilus* when their own vessel is sunk. Nemo, the lover of liberty, keeps them imprisoned because if they escape they might reveal his technological secrets. The high-tech environment of the submarine allows Nemo to escape from authority, but to defend that environment he himself turns into a jailer. Even worse, he becomes a murderer, sinking the *Abraham Lincoln* and another frigate when they chase after the *Nautilus*.

Verne recognizes these contradictions; he puts them in the mouths of the captives. Professor Aronnax berates Nemo for taking the lives of innocent sailors. He reminds Nemo that he and his two friends have been on board seven months and asks if they are going to be kept there forever; "You impose actual slavery on us!" Nemo, arms crossed, gazes intently at Aronnax and responds that he has never denied them the right to regain their liberty.[49] It is an unspoken challenge to escape—to repeat Nemo's own flight to freedom. Freedom, for Nemo, is defined as the creation of an artificial environment where, as he proclaims, "I am the law, and I am the judge!"[50] For others, though, Nemo's liberty means slavery or revolt.

In his lonely defiance, Nemo is in fact a romanticized, attractive version of Silfax—or, to state the matter more accurately, since *Twenty Thousand Leagues* was written first, Silfax is a degenerate version of Nemo. Both are rebels driven mad (or half-mad) by trials and dangers that have caused them to develop an implacable hatred for humankind. They plunge beneath the surface of the earth in order to evade the rules they despise. There, in darkness, they can hide from society and live by their own rules. (When the *Nautilus* arrives at the South Pole, Nemo unfurls the black flag of anarchism and proclaims himself king of the underworld: "Let a night of six months spread its shadows over my new domains!"[51]) And there, in the underworld, they become mysterious agents of doom, suddenly erupting to terrorize surface society. Both become destroyers and are finally themselves destroyed (although Nemo is first redeemed by becoming, for the beleaguered colonists of *The Mysterious Island,* a shadowy agent of good).

Verne's technological dream is clear and powerful. He is obsessed by the vision of a penetration of nature so profound and complete that the human world and the natural world become one. His political dream, though, is confused and shaky. He never wants to admit that power over nature entails, to some degree, power over people. Verne passionately believes that technology serves liberty, but it is the liberty of the runaway. He never resolves the problem of defending the technology-based free environment against external threats without repeating the destruction and injustices of the surface world.

These confusions, which are only in the background in *Twenty Thousand Leagues* and *Black Indies,* became more prominent in Verne's novels written between 1880 and 1890. In those later novels, the themes of destruction and irrationality became even more dominant. The ideal, enclosed community of labor becomes transmuted into an equally enclosed but now militarized, corrupt, and rigidly planned "city of perdition." Chesneaux remarks that at least one of these cities is a "vision worthy of that of Fritz Lang in *Metropolis.*"[52] It is also a vision worthy of H. G. Wells.

If Verne moved from optimism to pessimism, Wells seemed to move in the opposite direction. His early "scientific romances," such as *The Time Machine,* are deeply pessimistic visions of technological progress and ruling elites. But in his middle years (notably from the 1920s on), he wrote his now largely forgotten "Wellsian" works—propaganda for a highly mechanized World State run by a technological elite, the Samurai. Like Verne, Wells was a creature of contradictions. He encompassed "world planner and dystopian, artist and journalist, philosopher and propagandist, being and becoming, evolution and ethics, optimism and pessimism."[53]

The one constant in the works of both Verne and Wells is a fascination with modern science and technology. Wells, though, is much more conscious of the connection between scientific-technological power and political power. As we have seen, Verne downplays issues of political authority and class conflict. He prefers to imagine communities that act like one big family in cooperating to share the fruits of scientific knowledge. Wells has no such illusions about the possibilities of stateless, classless science.

Historical circumstances account for some of this difference in political perspective. Verne was writing in the idealistic days of France's Third Republic, when the prevailing ideology proclaimed technological progress and class harmony as inseparable virtues. Wells, on the other hand, was writing in Britain in the 1890s, in a depressed atmosphere of Social Darwinism and labor unrest. Another reason for the difference is more personal. If Verne was a scientific amateur, Wells was a professional. For Wells science was a vocation, a way to pull himself out of the lower middle class and to get a footing on the road to wealth and fame. In a very personal way, then, Wells experienced the liberating power of science. Yet he was also aware, both through his own pessimistic nature[54] and through Huxley's influence, that the power of science could be used in evil ways. Many of the contradictions in Wells's writing emerge from this conflict between his own beneficial experience with science and his awareness of its potentially pernicious effects on society.

If the key question for Verne is how to reconcile technology and liberty, for Wells it is who will control science and technology. He imagines a variety of possibilities, including the mad scientist, the sane scientist, the Samurai, the capitalist bosses, and the proletariat. Some of these possibilities seem pessimistic; others seem optimistic. In all cases, however, Wells assumes that some group will control science and technology, and that this control is the fundamental element of political power.

Wells wrote his three subterranean tales in the space of six years: *The Time Machine* (1895), *When the Sleeper Wakes* (1899),[55] and *The First Men in the Moon* (1901). If these three stories are read in chronological order, they reveal an important evolution in Wells's political thinking. *The Time Machine* presents a simple, two-tiered political universe where (contrary to first appearances) the tough proletarians are in charge. The giant vertical city of *Sleeper* is much more complicated, for it contains multiple levels of technology and therefore of political control. In that city, the tough ones who end up ruling the world are descended not from the workers but from the bosses. The proletarians attempt to revolt, but it is unlikely they will succeed. In the hollowed-out moon of the third story, the control of the elite is so absolute, and the degeneration of the workers so complete, that revolt is not merely unlikely but impossible. Wells begins by merging Darwinism and Marxism, by demonstrating how human evolution will confirm Marxist hopes; he ends by showing how evolution will dash Marxist hopes by leading to the fatal weakening of the working class.

All three stories are political fictions—more specifically, political mystery stories. In each the plot involves the quest of the narrator (or, in the case of *Sleeper,* of the awakened Graham) to find out who is in charge of the dominant technologies. In each case, what is hidden below the ground is not just a technological structure but also a political one.

In *The Time Machine,* the Time Traveler presents three separate theories to explain what he sees in the world of 802701 A.D. His first theory is that of "an automatic civilisation and a decadent humanity." Because he does not understand the details of the technical arrangements, he has only "a general impression of automatic organisation."[56] The key word *automatic* describes both the technology and the civilization. He assumes he has arrived in a society where something akin to vril or to the Machine has magically removed the need for labor.

The narrator has to revise this hypothesis when he first glimpses the whitish, nocturnal Morlocks. He then suggests a second set of theories, based on his new realization that the technology of this society is not magical, that the machines do not run themselves: "What so natural, then, as to assume that it was in this artificial underworld that such work as was necessary to the comfort of the daylight race was done? The notion was so plausible that I at once accepted it. . . ."[57] The Time Traveler extrapolates from the distinction, in his own time, between Capitalist and Laborer. The tendency to put "the less ornamental purposes of civilisation" underground, and to reserve the prettier countryside for the wealthier classes, must have continued until "above ground you must have the Haves, pursuing pleasure and comfort and beauty, and below ground the Have-nots; the Workers getting continually adapted to the conditions of their labour." The leisured classes, he now hypothesizes, have triumphed not only over nature but "over nature and the fellow-man."[58]

The third and final theory emerges only after the Time Traveler has actually journeyed into the Morlocks's subterranean lair in order to recover the Time Machine. Down there he discovers the pounding machinery, the bloody meat, the inhuman malignity. When he combines this evidence with his new awareness that the Eloi are terrified of dark nights around the new moon, the narrator revises his theories yet again: ". . . with sudden shiver, came the clear knowledge of what the meat I had seen might be."[59] After the Morlocks attack and kill Weena (the Eloi woman whom he has befriended), the Time Traveler finally solves the political mystery. The old order, in which the aristocracy ruled and the workers served its needs, had gradually been reversed. In the "great quiet" that had followed humankind's conquest of natural and social insecurities, the leisured elite had slowly degenerated, losing intelligence and character. The Time Traveler explains: "The Under-world being in contact with machinery, which, however perfect, still needs some little thought outside habit, had probably retained perforce rather more initiative, if less of every other human character, than the upper. And when other meat failed them, [the Morlocks] turned to what old habit had hitherto forbidden."[60]

The Morlocks' control of machinery is therefore decisive. They inherit the earth and get their revenge because their need to labor has kept them intellectually active, although morally degraded. The degeneracy of brutality triumphs over the degeneracy of feebleness.

In *When The Sleeper Wakes* the topography is much the same—pounding machines far underground, pleasure above—but the solution to the political mystery is quite different. Solving that mystery means life or death to the awakened Graham. He has to figure out who is friend and who is foe, who will imprison him and who will save him, when to follow and when to fight. Most of all, he has to figure out his own identity as the mysterious Sleeper.

The city of *Sleeper* is not a bipartite world, split into upper and lower halves, but a multi-tier world with nearly infinite gradations between the glass dome at the city's roof and the underground factories of the Labour Company. Graham's movements are up and down, beginning on the roof spaces of the city and moving down to the hall of the people, back up to the flying stages above the roof surfaces, back down to the laboring "under side" of the city, and back up to contest Ostrog for mastery of the earth. As he gradually solves the mystery of his own role, Graham repeatedly finds himself looking far up, or far down, in this great glass-domed city of super-imposed galleries, moving roadways, bridges, lifts, ledges, and tunnels. His first impression of the city is one of "overwhelming architecture":

The place into which he looked was an aisle of Titanic buildings, curving spaciously in either direction. Overhead mighty cantilevers sprang together across the huge width of the place, and a tracery of translucent material shut out the sky. Gigantic globes of cool white light shamed the pale sunbeams that filtered down through the girders and wires. Here and there a gossamer suspension bridge dotted with foot passengers flung across the chasm and the air was webbed with slender cables.[61]

This vertical decor has since been repeated in countless works of science fiction, not only in books (e.g. Orwell's *1984*) but also in countless films (Wells saw Fritz Lang's *Metropolis* and pointed out how closely the look and the plot of the movie resembled those of *When the Sleeper Wakes*).[62] What is special about Wells's version is his emphasis on the connection between technological and social sublimity. In the aesthetic of sublimity (whether applied to natural or to technological objects), the effect of scale is to overwhelm the viewer, to make him feel small and impotent in comparison with infinite vistas and grand perspectives. In the sublime city of the Sleeper, human beings are perceived as minuscule and unimportant. When Graham is gazing at the urban vista for the first time, "a remote and tiny figure of a man clad in pale blue arrest[s] his attention."[63] Human beings are either tiny figures in the distance or anonymous dots in a huge cheering swarm.

The key events of the novel are ones where Graham/the Sleeper is confronted with a chaotic crowd. He is cheered without having a clue as to why. Shortly after he awakens and finds himself imprisoned in some inner room in the city, an unnamed guide leads him through a maze of girders, wind-wheels, and beams to the edge of a vast black pit. Terrified, Graham falls upon a flying machine and literally leaps into the dark to escape. When he lands, he finds himself in "a brightly lit hall with a roaring multitude of people beneath his feet. The people! His people! . . ." The narrative continues: ". . . He saw no individuals, he was conscious of a froth of pink faces, of waving arms and garments, he felt the occult influence of a vast crowd pouring over him, buoying him up. There were balconies, galleries, great archways giving remoter perspectives, and everywhere people, a vast arena of people, densely packed and cheering."[64] This is only the first of many vivid images of social sublimity in *Sleeper*. The leap into the lap of the people culminates in the victory of Ostrog and in Graham's installation as Master of the Earth, owner of half the world. Graham gradually learns, however, that Ostrog is not a savior but another jailer; he only intends for Graham to be a puppet king, while he wields the real power behind the scenes. A mysterious woman named Helen Wotton manages to inform Graham that nearly a third of the population lives in virtual slavery as part of the Labour Company, and that these oppressed people look to him, the awakened Sleeper, for rescue.

His conscience aroused, Graham tells Ostrog that he wants to journey downward into the depths of the city to learn "how common people live—the labour people more especially—how they work, marry, bear children, die—." Ostrog replies, "You get that from our realistic novelists." "I want reality," says Graham, "not realism."[65] And so Graham makes his own journey of moral and political education. In disguise, he first samples the debased life of the city streets. Then he descends even lower, carried by elevators down into the "great and dusty galleries" of machinery, where men, women, and children dressed in the blue canvas uniforms of the Labour Company tend the machines. They have "pinched faces . . . feeble muscles . . . weary eyes," and some are disfigured by the chemicals used in their labor. The subterranean machine world resembles Piranesi's prisons: "Shrouded aisles of giant machines seemed plunged in gloom. . . . crowded vaults seen through clouds of dust . . . of ill-lit subterranean aisles of sleeping places, illimitable vistas of pin-point lights." All this Titanic architecture is "crushed beneath the vast weight of that complex city world."[66]

Graham's subterranean journey ends abruptly when the rumor starts that Ostrog has betrayed the people. Once again Graham finds himself in a scene of chaos, feeling passive and confused as the people cheer. At this point, as Ostrog escapes in an "aeropile" (a small aircraft with brightly colored sails, available only to the rich), Graham finally realizes that *he* can be in charge, not as a puppet but as a ruler. He prepares to make a proclamation to the People of the Earth, but is suddenly overcome by doubts: "Abruptly it was perfectly clear to him that this revolt against Ostrog was premature, foredoomed to failure, the impulse of passionate inadequacy against inevitable things. He thought of that swift flight of aeroplanes like the swoop of Fate towards him."[67]

What rescues Graham from this latest attack of confusion and paralysis is the return of Helen Wotton. She reminds him that he is Master still, and under her inspiration he declares his intention to give all he has to the people—either to live for them or to die. Still, as invading aeroplanes loyal to Ostrog come closer and closer, Graham's sense of passive fatalism returns. As he and Helen hear the battle reports, they seem unreal even to themselves: ". . . the only realities in being were first the city that throbbed and roared yonder in a belated frenzy of defence and secondly the aeroplanes hurling inexorably towards them over the round shoulder of the world."[68] Graham tells Helen that the sound of the aeroplanes make him feel "as if I were fighting the machinery of fate."[69] One last time, though, her presence helps him resolve to challenge fate. Graham decides to meet the oncoming aeroplanes in a smaller, nimbler aeropile. The story ends with an exciting description of an aerial dogfight (eerily predictive of World War I and beyond) in which Graham suffers yet another fall, this time as his plane spins toward the ground. He thinks of Helen ("She at least was real") and suddenly becomes aware "that the earth was very near."[70] Graham's final realities are Helen and the earth, not the city and airplanes—but he discovers those realities only at the moment of his death.

The technological environment of *Sleeper* is inherently political. It imbues Graham with a submissive attitude of fatalism before power much greater than his own. In a 1921 preface to the book, Wells describes it as a vision of "base servitude in hypertrophied cities," adding that "the great city of this story is no more then than a nightmare of Capitalism triumphant."[71] Yet the city *is* more than a nightmare symbol of capitalism; it is also an objective source of capitalism's power. Far more than the enclosed space of the Pantheon, the enclosed space of the city conveys the authority of the civilization. The authority of the city, at once architectural and political, is an important reason why Graham is so confused and unable to take decisive action, other than fleeing for his life. Just because the city is so huge, he has to explore upward and downward before he can understand the social totality. And once he does begin to understand, he finds himself numbed by the scale of his surroundings. After his first views of the city's "overwhelming architecture," Graham feels himself "a little figure, very small and ineffectual, pitifully conspicuous."[72] This same feeling keeps returning as he keeps losing his way amid vast buildings or equally vast crowds. As a result, he doubts his own ability to take charge. The human-made landscape, the city itself, and the airplanes are "the only realities." Those who are in control of those crucial structures—namely Ostrog and his clique—take on the same fated quality as the built environment.

Graham's death may ensure his spiritual salvation, but whether it brings salvation to the people is by no means certain. The disciplined Ostrog makes plans; the people, like waves in the ocean, do not. They are capable of brave resistance, but for the most part they are only a chaotic throng cheering their latest leader. With Graham's death, it is hard to imagine how the people will be able to carry through a successful revolution. As Ostrog tells Graham: "Liberty is within—not without. . . . Suppose—which is impossible—that these swarming yelping fools in blue get the upper hand of us, what then? They will only fall to other masters. . . . Let them revolt, let them win and kill me and my like. Others will arise—other masters. The end will be the same."[73] Graham murmurs, "I wonder" in response, but there is nothing in the story to discredit Ostrog's assessment.[74]

In *Sleeper* Wells discards the idea that simple contact with machinery will keep workers tough and intelligent. In *The Time Machine,* there is simply a technological level and a nontechnological one. In the multi-story city of *Sleeper,* there are multiple levels of technological control. Tending machines makes the workers weak, not strong. Anyone with any strength or vitality finds a way out of the underways; the rest are condemned to a lifetime of mind–numbing repetition, disease, and oppression.[75] Wells recognizes that the evolutionary tendency of technology is not necessarily to democratize but may well be to differentiate. Political control depends upon control of key technologies. In *Sleeper* these are technologies of the sky—wind vanes and aircraft—not technologies of the cellar. In *Sleeper* (in contrast to *The Time Machine*) those made cruel and strong by contact with machinery are not the laborers but the bosses— "wicked, able" men such as Ostrog.[76] Gone are the weakling descendents of the traditional aristocracy; the new rulers are the technocrats. "This," Ostrog proclaims, "is the second aristocracy. The real one. . . . The common man now is a helpless unit. In these days we have this great machine of the city, and an organisation complex beyond his understanding."[77]

Wells's conviction that those who manage key technologies will inherit the earth—or the moon—is even more pronounced in *The First Men in the Moon,* published just two years after *Sleeper.* Once again he begins with a simple whodunit: Who stole the sphere, enameled with gravity-defying Cavorite, that got the absent-minded scientist Cavor and his plucky companion Bedford to the moon? After the two travelers discover that the moon is populated, the mystery becomes more general, more political: Who rules the moon?

Once again, the crucial evidence lies underground. This time the sleuths must make two subterranean journeys to solve the mysteries. The first, an involuntary journey, is made by Bedford (the

narrator) and Cavor when the moon's inhabitants, the Selenites, drag
them down below. They find themselves in an immense cavern filled
with whirring thudding machinery and lighted by a cold blue phosphorescent fluid. Here they meet the Selenites (small antlike creatures
with tentacles and a cylindrical body-case) and their domestic beasts
(the huge, flabby, white mooncalves, "like stupendous slugs, huge,
greasy hulls . . . monsters of mere fatness,"[78] which bleat and bellow
as the Selenites jab at them with prods and goads). When the Selenites
try to goad Cavor and Bedford too into crossing a plank across a
black gulf, Bedford resists. A fight ensues, and Bedford and Cavor
flee toward the surface through "tremendously huge" blue-lit galleries and chasms ("I can scarcely hope to convey to you the Titanic
proportion of all that place—the Titanic effect of it"[79]).

Once back on the moon's surface, Bedford and Cavor separate
to search for the missing sphere. Bedford discovers it, but then he
cannot locate Cavor. Having returned to earth alone, he begins to
receive wireless messages from the moon. They have been sent by
Cavor, who explains that he was again attacked by the Selenites and
taken back into the lunar interior. On this second journey, Cavor
descends much further. What appear to be lunar craters are in fact
(Cavor relates) the lids to lunar shafts running almost a hundred
miles downward. The shafts connect with one another by transverse
tunnels and expand into great caverns. Toward the center of the
moon the blue phosphorescent light grows stronger and the air
currents more powerful; at the center of the moon is "a lake of
heatless fire . . . like luminous blue milk that is about to boil." The
moon's vertical structure is a mirror image of the earth's.

Cavor discovers that as he approaches the geographic center of
the moon, he also approaches its political center. The Selenites who
live near the surface, he learns, are just one type—relatively unimportant mooncalf herders, butchers, and the like. As Cavor descends
he sees that the Selenites come in many different forms. Some seem
all legs, others all nose; some have lobster-like claws; the ones in
charge have hugely distended brain cases. Each form is "exquisitely
adapted to the social need it meets."[80] "In the moon," says Cavor,
"every citizen knows his place. He is born to that place, and the
elaborate discipline of training and education and surgery he under-
goes fits him at last so completely to it that he has neither ideas nor
organs for any purpose beyond it."[81] A Selenite destined to be a

minder of mooncalves is trained to be wiry and active, to find diversion in mooncalf lore, and to take no interest in the deeper part of the moon. Selenites who do fine work are "amazingly dwarfed and neat." As with ants, the task of reproduction falls upon a special, small class: "the mothers of the moon-world, large and stately beings beautifully fitted to bear the larval Selenite."[82] Cavor is not quite sure how these exquisite adaptations are achieved, but he tells how he recently "came upon a number of young Selenites, confined in jars from which only the fore limbs protruded, who were being compressed to become machine-minders of a special sort." "That wretched-looking hand sticking out of its jars," he continues, "haunts me still, although, of course, it is really in the end a far more humane proceeding than our earthly method of leaving children to grow into human beings, and then making machines of them."[83] As a result of this hyperspecialization, the Selenites are helpless to revolt. Indeed, many of them are unable even to think about revolting.

The moon is ruled by the swollen heads, the chief of which is the Grand Lunar. At the climax of the tale, Cavor has an audience with this Master of the Moon. The hall in which the Grand Lunar presides is filled with light, incense, and receding colonnades; its splendor makes even Cavor "feel extremely shabby and unworthy."[84] The ruler himself has a brain case many yards across, nodding above a dwarfed body.[85] Alas, in response to the Grand Lunar's questions about the earth, the tactless Cavor forgets himself and talks about war, particularly about the way human beings glorify its irrationality and destructiveness. Rather than expose the moon to such insanity, the Selenites evidently put Cavor to death.

The First Men in the Moon was published eight years after Émile Durkheim's classic work *On the Division of Labor in Society* (1893), which celebrates the specialization of labor that develops with more complex technologies. Evidently, Wells looked at the political consequences of specialization and concluded that technological differentiation would lead to tyranny, allowing a tiny elite to rule by virtue of its cunning and discipline. Comparing *The Time Machine* against *Sleeper* and *First Men in the Moon* reveals a reversal in the outcome of the political mysteries. In *The Time Machine,* the narrator discovers that the workers have inherited the earth and that the privileged classes have degenerated into weaklings; in *First Men in the Moon,* the narrator learns that the workers have degenerated into the helpless victims of tyranny and that the descendents of the bosses rule.[86] Not those who work with machinery but those who understand how to control it have become the masters of the universe.

In this chapter and the preceding one, two journeys of liberation have been traced. The first is the journey made by a defiant individual to escape the oppression of the collectivity. Sometimes this is a retreat in the direction of nature—not the horizontal retreat from urban to pastoral surroundings typical of American literature, but a vertical retreat from the technological and social enclosure of the underworld toward surface nature. This is the path taken by Bulwer-Lytton's narrator, Tarde's lovers, Forster's Kuno, the captives aboard the *Nautilus,* and Wells's men in the moon. When the direction of movement is vertical rather than horizontal, the difficulty of the journey is considerably enhanced. Going to the surface will probably mean permanent rather than temporary exile; it can mean homelessness or even death. Thus the flight from the collectivity is more radical and uncompromising than the American flight from city to country, because there is less possibility of return. But no matter how radical (and this is also true of the American examples), the escape from the collectivity is politically ineffective. It may lead to an experience of fulfillment for the individual, but the society goes on functioning as before.[87]

In the second type of journey, the journey into the social depths, the traveler usually expects to return to the surface. He seeks out, or is forced to confront, subterranean social realities that are hidden from the surface-dwellers. This journey is intended as a temporary episode of moral education, not as a permanent self-exile. Again, though, the trip tends to be politically ineffective. After the traveler returns, then what? He may become more generous and charitable in his private life, like Dickens's gravedigger or Scrooge. He may even lead a revolution, as Graham does, but this too may well be hopeless. The social problem expressed in this journey is the split between the exploited and the exploiters. The psychic problem it expresses is the split between intellectual awareness and effective action.

Still, in each case the very existence of the story implies the possibility of establishing some connection between the surface realm and the subsurface one. These realms are split topographically, but they are joined by the narration. The more profound disjuncture lies between the two types of narration—the journey of self-discovery and that of social discovery. A narration that would combine a journey of personal liberation and one of social liberation is hard to imagine.

The stories discussed above present two types of tyranny: informal but crushing social authority and oppression of a submerged class by the armed state. These forms of oppression are not necessarily alternatives, for the social sublime can be simultaneously mind-numbing and unjust. But when liberation is split into two kinds of narrative journeys, the connection between personal and collective discontent is severed. The discovery (or rediscovery) of personal consciousness is one tale; the discovery of social consciousness is another. The project of personal change seems unrelated to social change, and the trials of the consumer are disconnected from those of the producer.[88] A story that would unite the two quests seems as self-contradictory as (to use Leo Marx's phrase) "revolutionary pastoralism": "It is not easy to imagine how an aesthetic ethos of disengagement and renunciation could be lastingly incorporated in an authentic revolutionary movement—one that aims to transform the system of wealth and power."[89]

Some of the narratives discussed here do hint at a synthesis. The Sleeper Graham travels both upward and downward in the vast glass-domed city, and he simultaneously figures out both his own identity and the power structure of the society. When Kuno reaches the dale of Wessex, he discovers not only unspoiled nature but also the Homeless who will inherit the earth after the Machine destroys itself. In both cases the development of physical prowess links personal and social consciousness, allowing these heroes to master their technological environment: Kuno learns to climb through the maze of tunnels, and Graham learns to fly. This mastery allows them to discover a social order more natural than the machine-based collectivity. Kuno discovers, to his surprise, that the hills of Wessex nurture a human-scale community. Graham discovers, in fierce battles in the city's subterranean passages (to quote T. J. Clark on Baudelaire's response to the revolutions of 1848) "Nature itself, appearing bloody and fertile in the heart of the city."[90] In such events the discovery of self, of nature, and of social reality occurs simultaneously. Still, both Graham and Kuno meet violent deaths, victims of technological catastrophe. They suggest but hardly affirm the possibility of synthesizing the personal quest and the social one.

The difficulty of making connections between these two quests is disturbing. More disturbing still is the possibility that neither quest will be undertaken at all. To remain a live form of literature, these fictional journeys must have some objective correlative in daily human experience. The recurrent theme of the pastoral retreat in American literature is linked to the "recurrent 'event' of the discovery and continuing resettlement of the continent."[91] The theme of an underworld society in British and French literature is linked to the recurrent event of excavation, which reshaped England and the Continent during the nineteenth century. It is also linked to the unavoidable sight of a new class of poor who seemed to inhabit subterranean worlds where mechanism had replaced nature—the worlds of city slums, of factory enclosures, of mining districts.

But sights like these are becoming less and less a part of ordinary experience. Production continues to be moved to the Third World, where it is invisible to consumers. Back home, cities have become radically split into poor central districts and rich surrounding ones, instead of being the chaotic mixtures of Hugo's and Dickens's day. As a result, the privileged consumer is less and less likely to be confronted with the sight of workers laboring in unhealthy sweatshops and factories. To be sure, anyone visiting a large city is still likely to see another type of "degenerate," the kind Hugo talks about in *Les Misérables*—unemployed, drifters, criminals, derelicts. These inhabitants of the social depths may be just as much victims of structural inequities as exploited workers. Because they are so strange and frightening, however, they are far less likely to prick the social conscience of the middle-class visitor. The more sympathetic poor tend not to be seen, while those who are seen tend not to be sympathetic. Even more than in the late nineteenth century, in our own time the journey into the social depths requires a sort of heroism. Today the fictional tale of descent into the social underworld has less and less resonance with ordinary life.

At the same time, the other journey of liberation—from an oppressive, highly mechanized society up toward the sunlit surface of nature—is also losing its real-life equivalent. One has to retreat further and further to find a country landscape unblighted by litter, and even further to find a semblance of wilderness. The real-life equivalent of this fictional trip is not impossible, but it is becoming ever more expensive. Living in nature used to be the lot of the poor, while only the rich could afford many of the benefits of civilization. As Wells so clearly recognized, in the industrial age the retreat "back to nature" has more and more become a privilege for the wealthy. The vertical city of *Sleeper* reifies the social structure, with a small elite at the top (the only ones privileged to view the sky and the sun regularly) and the Labour Company below, people deprived of light and air and scenery. To use the term of the economist Fred Hirsch, nature is becoming a "positional good"–in this case, quite literally.[92] As with all such goods, there is greater and greater competition for an inherently fixed supply. And as with all such goods, it is less and less available to ordinary people, and it is increasingly limited to a small, privileged class.

So both the fictional journey into the social depths and the fictional journey back to nature are in danger of losing their resonance with real-life experience. But another journey of escape is becoming more common: the retreat further into artifice, into a self-constructed technological environment. This is the route taken by Captain Nemo as he seeks liberty below the waves in the *Nautilus,* by Beckford in his creation of a fantastic haven at Fonthill, by the fictional magus Edison in his underground laboratory, and by Vashti in her high-tech cell. This fictional journey is echoed on all sides in contemporary society in the form of retreats into personal or collective environments of consumption—the artificial paradises of the shopping mall or of the media room, for example. This is not a journey up to surface nature or down to the social depths; it is a journey further inward, a retreat from technology into technology. It represents not an urge to gain knowledge of the world, either natural or social, but, in Gissing's words, a self-absorbed urge to "make a world within the world."

In that case, our world will be dimensionless. It would have neither the horizontal dimension of the landscape nor the vertical dimension of the cosmos. If we cannot imagine either escaping from the social sublime or penetrating into its mysteries, we may well be afflicted even more cruelly by *spleen,* "the price of artifice . . . the world's deadly mimicry of our own lassitude and despair." We may feel, in the words of Irving Howe, that "we may destroy our civilization, but we cannot escape it."[93] Or we may feel that the only way to escape it is to destroy it.

The Underground and the Quest for Security

Howe comments, somewhat cutely but quite accurately, that we children of modernism "learned our *abc*'s lisping 'alienation, bourgeoisie, catastrophe.'"[1] The subterranean stories we have been considering here are haunted by the specter of disaster. This is not surprising; the proliferation of catastrophic images in the literature of the late nineteenth century is well documented.[2] These images, moreover, were not limited to the printed page. Even before the invention of cinema, danger was a key ingredient of popular entertainments in music halls and amusement parks on both sides of the Atlantic.[3]

By 1900 the taste for the sublime was being democratized. The deliberate search for "pleasing terror" (Burke's phrase) was no longer restricted to wealthier travelers who could seek out Fingal's Cave, Alpine peaks, or German mines. Those who could not go to Pompeii could have Pompeii come to them in the form of an amusement-park spectacle. The hunger for disaster of the late nineteenth century is, among other things, an example of the "democratization of luxury" that is an outstanding cultural and economic fact of that era—the luxury in this case being sublime experiences once available only to the wealthier classes.[4]

Most cultural historians, however, have assumed that the sources of modern "cataclysmic consciousness"[5] lie far deeper than the eighteenth-century taste for sublimity. Many commentators look for its origins in the ancient tradition of apocalyptic prophecy that runs through Judeo-Christian, classical, Indian, and Chinese cultures.[6] This tradition, like so many other sacred ones (including that of the journey to the underworld), began to be secularized in the middle of the seventeenth century. Just how it was secularized is hotly debated. Modern literature, modern ideas of progress, and modern totalitarianism have all been interpreted as spilt eschatology.[7]

All these cultural theories share a primarily psychological understanding of historical causation. When we dig down to find the historical truth, their proponents agree, we find a relatively stable subterranean level of consciousness that from time to time erupts into surface history. That buried psychic reservoir may be defined in Jungian terms, as archetypal images of destruction, or in Freudian terms, as a perpetual, insoluble tension between individual drives (aggressive, pleasure-directed, or both) and the realistic demands of civilized life.[8] In any case, modern images of destruction represent the overflow of powerful, repressed, persistent emotions of destruction.

One of the best-known Freudian interpretations along these lines is George Steiner's analysis in his T. S. Eliot Memorial Lectures of 1970, published in 1971 as the book *In Bluebeard's Castle*. Steiner reminds us of Freud's thesis in *Civilization and Its Discontents* that extreme tensions arise when civilized manners are imposed upon central, unfulfilled human drives; he also reminds us of Freud's thesis of "an inescapable drive towards war" arising from aggressive instincts. According to Steiner, such instincts were repressed during the long peace of the nineteenth century. The great letdown after the excitement of the revolutionary and Napoleonic eras gave rise to an ennui that in turn engendered an "itch for chaos," a "nostalgia for disaster."[9] Steiner summarizes his argument in this way:

Whether the psychic mechanisms involved were universal or historically localized, one thing is plain: by ca. 1900 there was a terrible readiness, indeed a thirst for what Yeats was to call the "blood-dimmed tide." Outwardly brilliant and serene, la belle époque was menacingly overripe. Anarchic compulsions were coming to a critical pitch beneath the garden surface. Note the prophetic images of subterranean danger, of destructive agencies ready to rise from sewerage and cellar, that obsess the literary imagination from the time of Poe and Les Misérables *to Henry James's* Princess Casamassima.[10]

First, the presence of hidden but powerful "psychic mechanisms"; next, the rising to the surface of "prophetic images"; finally, actual events of destruction prophesied by those images—as an example of this three-step sequence, Steiner calls attention to the fantasies of ruined cities that were common by the mid-1800s. He calls these fantasies "a characteristic 'counterdream'" of that century of progress and finds them prophetic of actual twentieth-century disasters: "Exactly a hundred years later, these apocalyptic collages and imaginary drawings of the end of Pompeii were to be our photographs of Warsaw and Dresden. It needs no psychoanalysis to suggest how strong a part of wish-fulfillment there was in these nineteenth-century intimations."[11] Once again, it seems, modern culture is haunted by a remnant of the mythical: the ruined cities of World War II reify ancient fantasies of destruction. In such interpretations, technology plays a secondary role to psychology. Technology projects or expresses preexisting mythical structures.[12] By giving primacy to imagination rather than to tools, such theories have the virtue of challenging technological determinism. On the other hand, if they go to the opposite extreme of psychic determinism, they leave the realm of history altogether. The emphasis on buried psychic pressures implies historical fatalism. If these pressures are so powerful and yet so unconscious, they are immune to change—immune to history.

"From a psychological point of view," Susan Sontag writes, "the imagination of disaster does not greatly differ from one period in history to another. But from a political and moral point of view, it does."[13] The moral and political uniqueness of the modern imagination of disaster lies in the breakdown of the age-old distinction between nature as a source of hazard and technology as a source of safety.[14] According to the traditional understanding, normal nature, aided by technology, benefits humankind. Disaster emanates from abnormal nature, an unusual state of loss in which technology is overpowered by natural forces. Technology maximizes the benefits of nature (sunshine, rainfall, mineral deposits) and minimizes its harmful potential (droughts, high winds, floods). Since the industrial revolution, however, the construction of a primarily technological environment has overturned these assumptions. The category of disaster no longer implies nature's assault against technology's shelter. Indeed, the pervasive fear of the twentieth century is not what nature can do to us, but what we might do to it.

Subterranean images can express a progressive, positive view of technology. Digging below the surface of the earth can be a quest for scientific truth, for technological power, or for aesthetic perfection. But going underground can also be a quest for security, and in these cases subterranean technologies express regression and fear. Humanity's earliest constructions were burrows rather than buildings. The wish to return to the dark, enclosed safety of the womb is so primitive as to be premythic. "In Neolithic times," writes the geographer Yi-Fu Tuan, "the basic shelter was a round semisubterranean hut, a womblike enclosure that contrasted vividly with the space beyond."[15] For the primate ancestors of humankind, this kind of concave artificial shelter gave "the physical and (one might guess) psychological security of the cave."[16] In the prehistoric or "prearchitectural" treatment of space, life processes were allowed to function away from light, away from natural threats. To be close to the earth was to be safe.[17] Even when the hut emerged above ground, its rectangular walls still proclaimed a powerful contrast between interior safety and exterior threat. These structures incarnate the fundamental technological project of providing shelter from a hostile natural world.

The paradox is that the built environment can itself become a prime source of risk. Once again the mine serves as a model. The miner is dependent upon many complicated technological systems for drainage, ventilation, lighting, and structural support. All these systems must function correctly, and this requires a high degree of social organization. A high degree of psychic discipline is also required: the miner himself must be ever vigilant and highly trained in safety precautions. If he fails, or if any of the technological or social systems fail, the consequences are serious, even fatal. There are no easy escape routes. The result of a mistake could well be that most terrifying of all nightmares: being buried alive. Lewis Mumford reminds us how the miner's flickering light throws upon the walls "monstrous distortions of his pick or of his arm: shapes of fear." The "manufactured environment" of the mine is, among other things, an environment of fear.

In this respect too, the technological environment of the surface world has come to resemble the subsurface world of the miner. We are highly dependent upon complicated mechanical systems; we require a high degree of social and psychic discipline to keep them functioning correctly; we are aware of the serious consequences of error. The dangerous conditions of subterranean labor have been diffused throughout society at large. Our quest to build an ever-safer technological shelter from natural hazards has paradoxically led to a profound and pervasive sense of insecurity. The cycle is at once endless and obsessive.

This paradox is amply documented in the subterranean narratives of the nineteenth century. In most of them, the motive for going underground in the first place is to escape some natural disaster. That disaster may be an "abnormal" catastrophe (a flood), a "normal" one (the cooling of the sun), or an ecological one (resulting from human overuse of the planet's resources). Once underground, however, society proves vulnerable to other catastrophes—not natural disasters, in most cases, but social ones arising from humanity's inability to live harmoniously in an enclosed environment. In these subterranean stories, the concept of disaster is greatly expanded from the traditional definition of an abnormal natural event. These disasters are social as well as natural in origin; they arise from normal life as well as from abnormal events.

The redefinition of catastrophe began in nineteenth-century science. At the beginning of the century, the word *catastrophe* had a reasonably clear scientific meaning. The "catastrophists" were students of natural history (including Cuvier, Élie de Beaumont, and Buckland) who agreed that infrequent but massive paroxysms had interrupted the normal course of nature, and that this had been the predominant mode of geological change. The opposing camp—the gradualists or uniformitarians, led by Charles Lyell—did not deny the importance of floods, earthquakes, and eruptions; however, they argued that such episodes were strictly local in their effects, and that they had not occurred nor would they occur any more frequently or violently than in the present era.[18] Instead, the normal course of nature, with its ordinary phenomena such as erosion and slow uplift, was primarily responsible for shaping the globe over immensely long periods of time. For the uniformitarians, belief in global upheavals was unscientific at best; at worst it was an unreasoned defense of Biblical authority. Lyell even suggested that the catastrophists were moti-

vated by unconscious, atavistic fears—by remnants of the mythological, so to speak:

*The superstitions of a savage tribe are transmitted through all the progressive
stages of society, till they exert a powerful influence on the mind of the
philosopher. He may find, in the monuments of former changes on the earth's
surface, an apparent confirmation of tenets handed down through successive
generations, from the rude hunter, whose terrified imagination drew a false
picture of those awful visitations of floods and earthquakes, whereby the
whole earth as known to him was simultaneously devastated.*[19]

Yet even the unsuperstitious uniformitarians were, in their own way,
catastrophists. They too assumed that the earth's surface had been
"devastated," though not all parts of it simultaneously. They too
claimed that the planet had been radically reshaped, entire oceans
raised and mountains sunk, some landforms completely destroyed
and others created completely anew. The difference was that the
uniformitarians assumed such massive changes were, in a sense, part
of the regular functioning of nature, rather than intermittent, unusual
episodes. For the uniformitarians, normal nature was catastrophic.

As the nineteenth century progressed, the traditional distinction
between a tranquil, beneficial, normal state of nature and abormal,
harmful episodes of disaster became even more smudged. The discovery of the second law of thermodynamics seemed to prove that
first the sun and then the entire universe would gradually cool and
finally subside into frigid lifelessness. The entropic process would
be far slower and less evident than a universal deluge or upheaval.
(Hermann von Helmholtz estimated that the sun would cool in some
hundreds of thousands of years; later estimates claimed 10 million
years.) Still, the cooling of the sun would be irreversible and total.
The law of entropy seemed to be final proof that, over the long term,
normal nature is disastrous.[20]

In the subterranean narratives discussed here, natural disasters,
whether abnormal or normal, are the prime reason for taking refuge
underground. The paradigmatic example of such a disaster is, of
course, the Biblical Deluge, which in the nineteenth century was
often identified with a Cuverian *bouleversement*. In Bulwer-Lytton's
The Coming Race this seems to be the upheaval that drove the ancestors of the vril-ya below the surface. Long ago, the narrator tells us,
the earth was "in the throes and travail of transition from one form

of development to another, and subject to many violent revolutions of nature." During one such revolution, the upper world was subjected to floods, "not rapid, but gradual and uncontrollable, in which all, save a scanty remnant, were submerged and perished."[21] The narrator refuses to speculate whether this was the Biblical Deluge or "some earlier one contended for by geologists." In any case, it was a "great revulsion" that changed "the whole face of the earth," turning land into sea and sea into land. A band of survivors took refuge in caverns among the higher rocks; "wandering through these hollows, they lost sight of the upper world for ever." The narrator ends his description of these geological events by alluding to buried evidence of even older civilizations: "In the bowels of the inner earth even now, I was informed as a positive fact, might be discovered the remains of human habitation—habitation not in huts and caverns, but in vast cities whose ruins attest the civilization of races which flourished before the age of Noah, and are not be classified with those genera to which philosophy ascribes the use of flint and the ignorance of iron."[22]

Thus Bulwer-Lytton managed to combine Biblical accounts, modern geology, and a vague belief in Atlantis-like civilizations. The combination may seem improbable now, but it was not at all unusual in the later nineteenth century. At a time when the composition of the earth's interior was still mysterious and subject to conflicting scientific interpretations, all sorts of beliefs could be projected into subterranean regions, where science could neither confirm nor deny them. The myth of Atlantis—the sudden death of a high civilization—was an especially gripping parable in the age of progress. Furthermore, at a time when archaeological science seemed to be proving the accuracy of ancient myths (notably through the discovery of Homer's Troy), it was not implausible to speculate that the ruins of Atlantis might someday be found. In Doyle's *The Maracot Deep*, a trio of explorers discover descendants of the Atlanteans in a trench of the Atlantic Ocean. In H. Rider Haggard's *The Day the Earth Shook*, the descendants are found deep underground. In each of these books, a technology of "thought projection" allows the explorers actually to see the sublime spectacle of eruptions and floods that sunk the island eons before—a sort of cinematic version of Plato's story, with Cuvierian special effects.[23]

Jules Verne, too, combined veneration for modern science and fascination with Atlantis. In *Hatteras,* Dr. Clawbonny muses about the theory of the French astronomer Bailly that the descendants of the Atlanteans are to be found at the North Pole. One of the climactic moments of *Twenty Thousand Leagues* comes when Captain Nemo leads Professor Aronnax to view the submarine ruins of that once-mighty civilization. Even more significantly, Atlantis dominates Verne's *L'Éternel Adam,* which was probably written around 1900. This tale is set far in the future; its hero, the archaeologist-scholar Zartog Sofr Ai-Sr, seeks the truth about the distant past. He is obsessed with finding evidence for the continuous and irreversible ascent of biological and social life. Through extensive archaeological research (digging trenches, decoding manuscripts), the Zartog learns instead that the universe has moved in endless cycles of disaster and renewal. The dry land and the sea have changed places; Atlantis has repeatedly been submerged and lifted above the waves; humanity has gone from savagery to civilization and back and forth again. By the end of his life, Verne, who began the *Voyages extraordinaires* with a fervent belief in humanity's conquest of nature, seems to have concluded that nature would eventually reassert its ultimate authority.

The impending heat-death of the earth seemed to demonstrate even more powerfully that nature is inherently disastrous. Verne was well aware of theories of entropy; he put them into the mouths of two clear-sighted heroes, Cyrus Smith in *Mysterious Island* ("Someday our world will end, or rather animal and vegetable life will no longer be possible because of the intense chilling that it will undergo"[24]) and Captain Nemo ("The earth will some day be a cold body. It will become uninhabitable and will be as uninhabited as the moon"[25]). The closest Verne came to fictionalizing such a catastrophe is in his 1877 story *Hector Servadac* [also known as *Off on a Comet*], in which bits of the earth and an assortment of its inhabitants are swept into space by a comet that brushes the planet's surface. At first the cosmic travelers are kept warm in outer space by lava issuing from a volcano on the comet, but when the lava stops flowing the temperature begins to plunge. The intrepid Captain Servadac persuades his comrades to descend into the heart of the volcano, where for months they endure "a life of dreary monotony,"[26] getting on one another's nerves, sleeping a lot, and succumbing to severe depression. When the comet begins to head back toward the earth, a thaw sets in and the travelers joyfully return to their planetary home via balloon.

There are so many points of similarity between *Servadac* and Tarde's *Underground Man* (the brave leader exhorting his numbed followers; getting food from frozen carcasses and water from glaciers) that we must assume Tarde had read Verne's book. Tarde's own account of entropic cooling is one of the more lighthearted descriptions of catastrophe, for he delights in showing how nature mocks the so-called scientific and technological progress of the century. At first, scholars "in their well-warmed studies" dismiss the falling temperature as a short-term phenomenon. As things get worse, they write sensational, lucrative pamphlets on the subject. When "solar anemia" turns to "solar apoplexy," a global freeze sets in. Long trains are caught in Alpine tunnels, the entrances to which have been blocked by avalanches. Telegraph and electrical lines are buried in snowdrifts. As glaciers advance south, the triumphs of the century are buried by a moving cliff in which are embedded machines and buildings, bridges and boulders, all heaped together in a frozen jumble, all "whirled along in the wildest confusion, a heart-breaking welter of gigantic bric-a-brac."[27]

The theme of going underground to escape the death chill of the planet is repeated in other popular novels of the period, notably Camille Flammarion's *La fin du monde* (1893) and William Hope Hodgson's *The Night Land* (1912).[28] H. G. Wells used the idea in "The Man of the Year Million" (1893), in which, deep within the earth, a band of supermen with huge brains bravely resist the encroaching cold as the sun dies out.[29] Similarly, in one of Wells's best-known works, *The War of the Worlds* (1898), the Martians invade the earth because their planet has grown too harsh to support life. In all these visions, it is the normal functioning of nature, not some unusual event, that threatens survival.

Whether the threat is an abnormal upheaval or normal cooling, human beings have no part in the origins of these disasters. Other subterranean narratives, however, tell of ecological disasters in which nature's failure is primarily caused by excessive human demands. The failure may be a form of pollution or the depletion of necessary resources; the cause may be overpopulation, overconsumption, or both. Ecological disasters are not accidents in the sense of unusual, harmful technological episodes. Just as the line between normal and abnormal nature was becoming blurred, so was that between normal and abnormal technology. Ecological disasters are normal developments when technological progress is defined primarily as growth in population and consumption.

In Hay's *Three Hundred Years Hence,* for example, the population has grown so prodigiously that humanity has had to burrow underground to find resources and space. While Hay praises the various technical feats that have so greatly increased the planet's carrying capacity, the dominant mood of the book is one of anxiety. It shows a vastly overcrowded planet, wholly dependent on complex technologies and ripe for a Malthusian crash. Just a year after the appearance of *Three Hundred Years Hence,* Hay published a smaller book that even more explicitly predicts an environmental disaster resulting from population pressures. *The Doom of the Great City* (1882) is supposedly a narrative written in 1942 by one of the few survivors of a killer fog that had struck London in 1882. The author tells how London had swollen to monstrous proportions. As the population grew, so did the severity of London's fogs, which seemed to descend more frequently and to cause more physical maladies. In 1882 a "portentous gloom" had descended upon the city, cutting it off from the sun and enclosing it in an envelope of darkness. This ecocatastrophe showed "the hand sublime of Nature, the supremacy of offended God." It mocked human pride and taught men "their own immeasurable littleness": "like the sudden overflow of Vesuvius upon the towns below . . . or like that yet older tale, a world had sunk beneath the waters, so, in like manner, the fog had drawn over midnight London an envelope of murky death, within whose awful fold all that had life had died."[30] The narrator wanders through the silent streets of London, viewing the corpses of the rich and the poor heaped together, all social divisions having been leveled by mass death.[31]

The prospect of ecological failure also provides the mainspring of the plot of Verne's *Black Indies.* In this case, though, the ecological catastrophe is depletion rather than pollution. *Black Indies* was written at a time of considerable anxiety about coal reserves. Beginning in the 1860s, analysts on the Continent and in Britain started to worry when they compared estimates of future consumption with those of future supply. William Stanley Jevons, the British economist, warned in *The Coal Question* (1866) that in little over a century British coal would be prohibitively expensive to mine. As a result, during the 1880s and the 1890s there was a great deal of interest in alternatives to fossil fuels—especially in hydroelectric power, which in France became known as *houille blanche* (white coal).

The plot of *Black Indies,* already discussed in chapter 6, arises from one miner's refusal to believe that the Aberfoyle pit is exhausted. Early in the book, young Harry Ford exclaims to James Starr, "It's a pity that all the globe was not made of coal; then there would have been enough to last millions of years!" The older, wiser engineer responds that nature showed more forethought than human beings by forming the earth mainly of stones that cannot be burned. Otherwise, Starr comments, "the earth would have passed to the last bit into the furnaces of engines, machines, steamers, gas factories; certainly, that would have been the end of our world one fine day!"[32] Verne was a fervent believer in nature's bounty and in mankind's ingenuity, but even more did he believe in respecting the facts of nature. James Starr ends his remarks on coal depletion by commenting: ". . . in my opinion England is very wrong in exchanging her fuel for the gold of other nations! . . . Unfortunately man cannot produce [coal] at will. Though our external forests grow incessantly under the influence of heat and water, our subterranean forests will not be reproduced, and if they were, the globe would never be in the state necessary to make them into coal."[33]

So far we have considered fictional disasters that drive humanity underground. For the most part, these disasters are natural, involving an episode of abnormal nature, a disaster inherent in nature's longer-term normal functioning, or an ecological catastrophe in which nature collapses because of excessive human demands. But there are also stories in which disaster strikes the subterranean refuge. In these stories, the disaster is more likely to be technological or social than natural.

To be sure, both Wells and Verne wrote some of their best passages about natural catastrophes that destroy surface and subsurface worlds alike. Verne's *Nautilus* survives the perils of the Maelstrom at the end of *Twenty Thousand Leagues*; however, in *Mysterious Island,* soon after the vessel is reverently sunk as a final resting place for Captain Nemo, the entire island (and presumably whatever is left of Nemo's ark) is blown up in a volcanic eruption. Verne's descriptions of the slow awakening of the volcano and of the settlers' reactions to impending doom are masterly. As for Wells, the most unforgettable passages of *The Time Machine* come at the end when the Time Traveler is trying to escape the pursuing Morlocks. He leaps onto his machine and mistakenly turns the wrong levers. Instead of returning to the nineteenth century, he spins even further into the future, where he witnesses the entropic death of the universe. Neither the Eloi nor the Morlocks survive as the earth inexorably chills. The sun becomes redder and cooler, the moon disappears, the air thins out. Finally, 30 million years hence, the Time Traveler comes to rest on the dying earth. The enormous sun obscures a tenth of the sky but fails to warm the earth, now locked in bitter cold. The landscape is one of rocks and stars, an utterly black sky, an absolute silence. This is Baudelaire's "Rêve parisien," a black, motionless world without vegetation, now a vision of death—the death of nature itself, except for one living creature: "a round thing . . . tentacles trailed down from it; it seemed black against the weltering blood-red water, and it was hopping fitfully about."[34]

In most cases, though, the disasters that destroy the subterranean societies are of human origin. Sometimes they resemble technological accidents. The case of the android Hadaly in *L'Ève future* is ambiguous, for the shipboard fire that "kills" her lies somewhere between natural and technological causes. The final catastrophe in "The Machine Stops," however, is clearly technological. The event resembles an earthquake, but it is in fact caused by an out-of-control aircraft that plunges into the underworld. This crash is just the death blow, however. The beehive society has already been disintegrating through technological entropy, a series of small breakdowns whose cumulative effect has been devastating. "The Machine Stops" is a story of the disasters that can ensue from deferred or inadequate maintenance: dispensers that no longer dispense, levers that fail to deliver, brown and smelly water, brownouts and glitches and interruptions in service. At first they are simply annoying, but they eventually destroy the social structure.

However, the most common type of disaster that destroys subterranean societies is not mechanical malfunction but social malfunction: war, revolution, terrorism. Of course, the threat of military attack may be a reason for going underground in the first place. In

Les Misérables, Jean Valjean descends into the sewers with Marius to escape the fall of the barricade. In *The War of the Worlds,* the narrator and a few others take shelter in a scullery, and later the narrator meets an artilleryman who has even grander notions of going below to escape the invaders: "You see, how I mean to live is underground. I've been thinking about the drains. . . . Under this London are miles and miles—hundreds of miles—and a few days' rain and London empty will leave them sweet and clean. The main drains are big enough and airy enough for any one. Then there's cellars, vaults, stores, from which bolting passages may be made to the drains. And the railway tunnels and subways. Eh? You begin to see?"[35] But in most of the narratives discussed here, the underground is less a refuge from war than a scene of war. The members of *The Coming Race* use bolts of vril to reduce any opponent—another underground tribe, a monstrous reptile, and potentially the narrator—to a heap of cinders. Among the neo-Troglodytes of *Underground Man,* Tarde tells us (though he spares us the details) there have been horrendous subterranean wars between the consumer and producer cities, and later between the federalist and centralist ones: "What human ear, nose, or stomach could have longer withstood the deafening roar and smoke of melanite explosions beneath our crypts; the sight and stench of mangled bodies piled up within our narrow confines? Hideous and odious, revolting beyond all expression, the underground war finished by becoming impossible."[36] Less delicately, Wells describes how the underground passageways of the giant city in *Sleeper* are slippery with blood and littered with bodies, as the People first battle for Ostrog's clique and then rise in general revolt. The book ends (somewhat like Forster's story) with an airplane plunging into the earth and killing the hero. In *First Men in the Moon* Wells describes a surreal battle scene in which Bedford strikes one of the Selenites and, to his surprise, sees the moon creature collapse from the blow like a thin, pulpy plant. The Selenites fight back and chase the two earthlings, who run in confusion among the blue-lighted caverns.

Verne is equally obsessed with subsurface (or submarine) warfare. Captain Nemo shoots at pursuing vessels and, in one case, leaves a crew to perish in the frigid ocean. Aronnax is ashamed to find himself fascinated by this spectacle of disaster, watching with "an irresistible attraction" as a "human ant-heap" of anguished sailors are "overtaken by the sea."[37] The climactic moments of *Black Indies* come when a loch empties into the underground cavern and when a firedamp explosion is narrowly averted. These are not acts of God, as they seem at first, but acts of a saboteur: the half-crazed Silfax.[38]

The reason for constructing an underground environment in the first place is to find security from nature's risks and limits. Fear of disaster impels human beings to seek more control over their environment, but lack of self-control leads to other fears and to other disasters.

Let us return to the problem raised at the outset of this chapter. Just how does the modern imagination of disaster differ, politically and morally, from its predecessors? More specifically, how are those imaginative differences related to the transformation of the human environment from a predominantly natural one to a predominantly technological one?

Such questions are addressed in Michael Barkun's thoughtful study *Disaster and the Millennium* (1974). Barkun establishes four factors as crucial to the emergence of traditional millenarianism: a rural, isolated setting inhabited by a relatively homogeneous population; multiple episodes of natural or social disorder; the presence of a potential leader; and the availability of a millenarian doctrine. In describing the second of these factors, Barkun emphasizes the distinction between accidents and disasters. Accidents are episodes that may severely affect a community but that do not lead to fundamental social change. Disasters, on the other hand, are repeated shocks that result in the collapse of what has variously been called the "true" society or the "mazeway"—"the perceived primary environment, that combination of people and things that an individual regards as inseparably linked to his own sense of meaning and well-being."[39] Accidents temporarily damage this primary environment; disasters (rampant overpopulation, prolonged scarcity, profound social changes) transform it entirely.

Barkun goes on to consider how the millenarian mentality might emerge in urban, technological settings rather than isolated rural ones. Because of modern media of communication, awareness of disaster can be shared vicariously even by those who are not directly affected, so that local events can have global impact. On the other hand, "institutions of concentration" (such as factories, universities, and even slums) have the effect of separating their members from society at large, creating homogeneous populations that are the modern urban equivalent of rural villages. Barkun concludes that those who have promoted the ideal of permanent revolution—the Soviet revolutionaries, the Nazis, the Chinese Communists—have exploited both these conditions to induce disaster at will in order to keep the revolution going. "Once the unintended consequence of disaster, millenarian expectation now flourishes as its cause."[40]

The ideology of progress is also an ideal of permanent revolution. As Barkun points out in referring to Ernest Tuveson's thesis, the idea of progress is a secularized, diffused form of millenarianism: "The notion of divine intervention in history gradually became the redemption of man's condition through science and technology, in a word, *progress*."[41] Neither the belief nor the believers are sinister and calculating; they represent some of humankind's best hopes, not its worst instincts. Still, in a far more muted way, millenarian expectations of progress can have catastrophic consequences. Not technology itself but what the nineteenth century understood as technological progress (an important distinction) can assume the aura of disaster.

Disaster here must be understood as Barkun uses the term: as a long-term, fundamental destruction of the primary environment. It is not just that progress inevitably brings destruction as an unfortunate side effect. Progress *is* destruction. It repeatedly and radically tears down established material and social forms. It reshapes the physical environment, not just once but over and over. Hills are blasted through, Stagg's Garden is destroyed by an "earthquake," the streets of Paris are torn up for years on end, familiar landmarks are cut through to make way for railway lines.

As the physical environment is undermined and transformed, so is the social environment. A general rise in the standard of living, however desirable in many respects, causes an upheaval in existing social structures. Old scarcities disappear, but new ones are created as expectations change and as capitalists repeatedly create new markets. Longer life expectancy is another effect of technological progress that, however desirable, creates a permanent state of social instability. The combination of higher population and higher living standards leads to pollution of some natural resources and depletion of others. In all these ways, technological progress tends to break down the "true" society or "mazeway."[42] It subjects the physical and social environment to profound, repeated shocks. In the words of Bertolt Brecht, "normal everyday life turns out to have become abnormal in a way that affects us all."[43]

If (as Brecht recommends) we shake up our ideas of normalcy and abnormalcy, we begin to understand why the imagination of disaster flourished toward the end of the nineteenth century. At a time when normal everyday life was proving to be abnormal in such a pervasive way, when the disastrous quality of technological progress was becoming more evident, a *real* catastrophe could have the paradoxical effect of restoring a more normal, natural way of life. Such an event would halt the degeneration of the individual, of the community, and of the physical environment. In all these areas, natural strength and health would be restored. What Steiner describes as "nostalgia for disaster" is better understood as nostalgia for nature.

A catastrophe would halt individual degeneration, in the first place, because only the strong would survive. The fear of cosmic entropy was a naturalized version of the dread of human degeneration, and vice versa. Both were seen as long-term, quiet, subversive sources of disaster. They were "natural" enough, given the physical order of the universe, but they were fundamental threats to human life as we know it. A disaster calls forth the heroic Miltiades, Servadac, or Jean Valjean. At the same time, it weeds out degenerate weaklings. In *The War of the Worlds* the artilleryman imagines that those who take refuge in the drains will form a band of "able-bodied, clean-minded" men and women. No rubbish, no cowards, no silly ladies will be allowed in. "Life is real again, and the useless and cumbersome and mischievous have to die."[44] The artilleryman illustrates Orwell's conviction, expressed in *Wigan Pier,* that "many of the qualities we admire in human beings can only function in opposition to some kind of disaster, pain or difficulty."[45]

While they strengthen individuals, crises also strengthen communities. There is no contradiction here, if one understands true humanity as including the capacities for loyalty and fellow-feeling. Barkun reminds us that a catastrophe is often followed by the emergence of a "disaster utopia" when the sense of common purpose leads to a flowering of good will, generosity, and altruism.[46] In this euphoric state, conventional social distinctions are swept away and are replaced by more egalitarian, democratic relationships. Communal bonds that have been weakened by a rising population and rising expectations are suddenly and dramatically restored. By removing prosperity and uniting people in fear, a catastrophe recreates the social bonds that erode under normal progressive conditions. It is an external agency that restores a lost community.

Catastrophe, finally, enables humankind to return to nature in the most fundamental sense of its everyday physical environment. With a few notable exceptions (William Morris's *News from Nowhere* is the most notable), the only way that modern writers can imagine a bucolic future of fields, daisies, and chipmunks is as the aftermath of some dreadful catastrophe. Barring that event, the environmental direction in which we are now headed seems to lead inevitably to a clogged, polluted, multilayer cityscape like that of Wells's *Sleeper*. It seems inconceivable that future humanity could dwell in a natural environment—not a primitive one, to be sure, but one where experiences such as seeing the starry sky, picking berries, and walking by an unlittered brook are commonplace. Such an environment is imaginable only if catastrophe, conceived of as an outside force, makes us change direction.

Nature is political. As we have already seen, the established political order recognizes physical nature as a source of subversion and ruins it for ideological as well as for practical reasons. A catastrophe would reassert nature's power against that order. A disaster would introduce an alternative source of authority: the authority of nature. Only then can one imagine the restoration of a more natural landscape. Only then can one imagine the restoration of a more natural community, as opposed to a radically split two-tier society or a dehumanized collectivity.

Catastrophe, then, means revolution. It is one of the few forms of revolution that still seem possible. Chiliastic movements are often interpreted as "prepolitical" or "archaic" forerunners of more rational forms of political protest, such as nationalistic movements, interest groups, and political parties. In the well-known phrase of Eric Hobsbawm, millenarianists are "primitive rebels."[47] But the yearning for destruction does not only represent a premodern revolutionary consciousness; it can also be interpreted as a *postrevolutionary* consciousness. At a time when the possibility of successful insurrection seemed over, as it did post-1871 Europe—when the triumphant march of technological progress seemed unstoppable and yet highly unstable—in this situation of frustration, of rebellious sentiments that had no outlet in action, fantasies of catastrophe became a form of political protest. They expressed political despair. These fantasies of destruction are therefore the flip side of fantasies of paradise. In both cases, a radical change is sought through supernatural means. In the one case, technological magic can create paradise; in the other, disaster is the magical way of removing an oppressive technological order that otherwise seems immovable.

In this book we have moved between subterranean imagination and subterranean reality many times, evaluating each in the light of the other. Let us make this move once more.

In the subterranean history of the nineteenth century, there is a striking disparity between actuality and imagination. In actuality, the major reason for excavation was progress, defined as the advancement of knowledge and the conquest of external nature. The bore beneath Mt. Cenis, the "clean, neat" sewers of Paris, the tunnel under the Thames, the trenches of Troy—all are bold images of scientific and technological advancement. In the imaginative literature of the century, however, the reasons for excavation are far more somber. Here the dominant motive for going underground is to find security from a natural or a social peril, or from some mixture of the two. These stories are therefore prophetic not so much of the disastrous events of our century as of our preoccupation with security. They foresee our enormous investment in subterranean projects motivated primarily by our desire to evade ecological and military disasters.

Certainly much of the twentieth century's excavation has also been motivated by goals of scientific and technological progress. The long-dreamed-of "Chunnel" is at last being constructed to provide a railway link between Britain and France. In the past several decades, cities like Boston and Paris have extended and modernized their subway networks. (Boston is also planning to excavate another harbor tunnel and to move a major highway underground.) In their quest to extend scientific knowledge, physicists have supervised the construction of a huge tank of highly purified water within the Morton-Thiokol salt mine underneath suburban Cleveland. Placed 1,000 feet underground to shield it from cosmic radiation, and approximately the size of a 6-story building, the tank is surrounded by electronic eyes programmed to capture the decay of a proton or a neutron.[48] The Stanford Linear Accelerator Center, south of San Francisco, includes a 2-mile-long tunnel, 11 feet high and 25 feet underground, in which the accelerator itself (a copper pipe 4 inches in diameter) is located.[49] The Superconducting Supercollider, to be built in Waxahachie, Texas, will include a tunnel about 53 miles in circumference.[50]

But far more than in the nineteenth century, subterranean projects in this century are motivated by nightmares rather than dreams. One major twentieth-century project arises from an ecological anxiety little foreseen in the nineteenth century: the need to dispose of nuclear wastes (much of it, appropriately, from plutonium). Enormous subterranean containers have also been used to address the problem of scarce resources. Underground caverns in Louisiana and Texas are used for the Strategic Petroleum Reserve of the United States (one in Louisiana, filled with oil, is about twice as high as the Empire State Building). Underground oil storage is also used extensively in Sweden, and half of Sweden's electric power comes from underground hydroelectric plants.[51]

The rising costs and uncertainties of energy supplies have motivated the construction of energy-efficient subterranean houses. A March 1977 *Popular Mechanics* article on underground architecture brought the largest mail response in that magazine's history. Architects such as Ken Labs and Malcolm Wells and organizations such as the Underground Space Center at the University of Minnesota have generated research projects, prototypes, and general enthusiasm for underground building. (In 1978 the Center prepared a booklet *Earth Sheltered Housing;* by the middle of 1979, the book had sold close to 70,000 copies.[52]) Commercial clients have also invested in subterranean buildings. In the United States the most extensive commercial use of underground space is found in Kansas City, where some 2,000 employees descend to workplaces created from abandoned limestone quarries.[53] The space costs only one-third to one-half as much as space in a surface building; heating and air conditioning cost about one-fourth what they would on the surface.[54]

For other urban planners and architects, the appeal of underground construction is as much ideological as practical. For them, efficiency is a subsidiary goal, while preservation of environmental quality in a crowded world is the primary concern. The French architect Édouard Utudjian, who in 1933 helped found the Groupe d'Études et de Coordination de l'Urbanisme Souterrain, distinguishes *habitat souterrain* (underground shelter) from *urbanisme souterrain* (underground urbanism). The former he considers an escape, though sensible in extreme climates; the latter he considers indispensable, believing that modern cities must extend in three dimensions to be livable.[55] In the Wellsian spirit of putting the less glamorous aspects of civilization underground, the architect Gunnar Birkerts has proposed citywide systems of underground conduits. These elaborate grids of trenches, 1,000 feet wide and 200 feet deep, would hold utility and distribution networks, central heating and cooling facilities, recycling and waste-treatment plants, and factories. Then, says Birkerts, the surface would be liberated for parks, housing, schools, and other less utilitarian functions.[56]

More in the spirit of William Delisle Hay, the Greek planner Constantinos Doxiadis foresees a crisis of overpopulation. He predicts a huge population growth that will lead to the universal city of Ecumenopolis, an agglomeration of 40 or 50 billion people surrounding islands of agricultural land and bits of open space. Stretched too thin to be civilized, Ecumenopolis will be a modern version of barbarism.[57] To avoid this disaster, humanity must instead plan Entopia, "the opposite of the many Dystopias in which we now live."[58] There the land surface of the city will be given back to the people for recreation and pleasure; factories and roads will be fully automated and sunk below the surface. "The few technicians who operate these underground factories [will] work in the buildings to be seen on the green open spaces."[59] Machines and utility below, nature and leisure above: the Morlocks and the Eloi.

An even more radically vertical concept of urbanity has been proposed by Paolo Soleri, again with the idea of preserving the natural skin of the earth against rampant population growth. Soleri's "arcologies" (from *architecture* and *ecology*) are three-dimensional cities, self-contained, highly complex, highly efficient, and miniaturized.[60] Soleri proposes a sort of urban implosion so that the unlivable, flat, spread-out city ("death by affluence"[61]) collapses into a superdense object, a sort of urban sculpture. Through arcology, Soleri believes, "man can, in fact, achieve an artificial landscape that vies successfully with the natural."[62] Soleri has experimented with "earth forming" as a building method, casting concrete in molds cut from the ground and using them to construct semisubterranean "earth

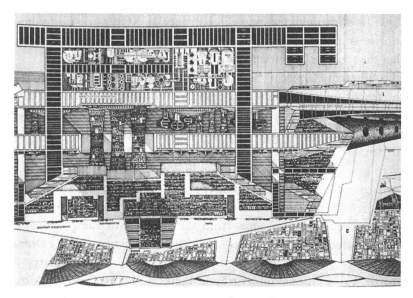

"Infrababel," one of thirty imaginary cities described by Paolo Soleri in *Arcology* (1969). Built in a stone quarry between two cliffs, Infrababel would have a population of 100,000 and would cover 148 surface acres while extending 400 meters vertically. Most of Soleri's arcologies are not completely underground but mimic subterranean conditions in their self-enclosed verticality. Courtesy of The MIT Press.

houses." These structures evoke the underground burrows described by Tarde in *Underground Man,* just as Soleri's social ideals echo Tarde's. (According to Soleri, "Human destiny . . . lies in the frugality demanded by the quest for essentials."[63])

Some of these grand plans for underground cities may be realized early in the twenty-first century. In Japan, where there are already too many people on too little land, task forces have been appointed to plan subterranean cities with populations ranging from 100,000 to 500,000 inhabitants, to be built in the early 2000s.[64] In this century, however, it is the quest for national security through military means that has prompted the most extensive and expensive subterranean construction. Tarde's description of a horrifying underground war—"the sight and stench of mangled bodies piled up in . . . narrow confines"—was realized in the trench warfare of World War I. The 25,000-mile network of trenches on the Western Front was a virtual underground city, a "troglodyte world" of level upon level, sometimes extending as deep as 30 feet, a world where sunrise and sunset were the only signs of nature.[65] Mumford said that the trenches seemed to turn Europe into a vast mine.[66]

When the Germans began to conduct zeppelin raids on London, the age of underground air-raid shelters also began. London's tube stations were hastily converted to bomb shelters. In World War II the network of air-raid shelters became far more extensive, especially in Britain. Unused mines were taken over by local authorities, as were caves around Dover. In London around eighty tube stations were used for refuge, along with constructed but unopened tunnels and stations. They sheltered not only people but also the Elgin Marbles and other treasures from the British Museum, as well as paintings from the Tate Gallery and Royal Academy. Other cultural treasures, including the crown jewels, were stored in a quarry in North Wales. Special underground bunkers were set aside for government functions, ranging from Churchill's "war room" to post offices to the BBC's emergency studios.[67] The RAF Bomber Command was directed from an underground operations room, constructed at High Wycombe in 1940.

All this now seems but a preview of the subterranean technologies of the nuclear age. Missile silos, buried command centers (such as the NORAD center in Colorado), bomb shelters large and small, and buried sites for vital government operations have occasioned stupendous excavation projects. In many cases the locations and other details are highly classified. One of the most public systems of shelters is that of Switzerland, which has openly adopted a policy of providing underground shelter space for most of its citizens. Among the largest of the Swiss shelters is the Sonnenberg Tunnel, a major roadway which in three days can be sealed off to provide blast and radiation protection for 21,000 people. This shelter includes an independent water supply, a two-week supply of fuel for electrical generators, a jail for twelve prisoners, survival rations, soothing music, and a 328-bed hospital.[68]

In our century the quests for military and ecological security have converged. The nineteenth-century fear that entropy would chill the planet has been transformed into the twentieth-century fear of a "nuclear winter." The ideal shelter would simultaneously guard against all terminal threats to the natural world. In quest of this ideal, a conservative Texan, Edward P. Bass, has provided most of the

The Operations Room of the Royal Air Force Bomber Command. This underground "nerve center" of the RAF, which went into use in March 1940, was protected by layers of concrete under grass-covered mounds. Its location in High Wycombe, Buckinghamshire, remained secret until several years after the war. Courtesy of John Laing plc.

development money (about $30 million) to build Biosphere II, a two-and-a-half acre enclosure that provides a rough model of the earth's ecosystems, ranging from desert to tropical rain forest. In December 1989 eight researchers plan to enter Biosphere II, where they will try to stay for two years, living off what it produces. "The grandiose goal of the venture is to prepare for a day when Earth might be no longer able to support life—because of the collapse of the sun, perhaps, or, more immediately, a nuclear war."[69] If the experiment is successful, Space Biospheres Ventures hopes to turn a profit by providing biospheres to research laboratories (for ecological experiments) and to individuals and organizations (as refuges from a "nuclear winter").[70]

Paul Fussell has said of the trenches of the Great War: "Thus the drift of modern history domesticates the fantastic and normalizes the unspeakable."[71] When so much money and attention is invested in subterranean shelters against future disaster, the present becomes distorted. When such projects are part of normal life, that life has become abnormal in a way that affects us all. A striking example is the enormous investment already made in the Strategic Defense Initiative, the ultimate air-raid shelter. By creating a new, supposedly invulnerable surface above us, SDI would have the effect of sheltering the entire nation underground.

Given our disastrous preoccupation with security, the most prophetic underground dweller may be the nameless furry animal of Franz Kafka's story "The Burrow." This story of an animal who in vain tries to elude an unseen but omnipresent enemy can certainly be read as Kafka's fruitless effort to elude death (it was written in the winter of 1923–24, the last year of his life, when he was suffering terribly from tuberculosis), but it can also be read as a commentary on the vain quest for safety through technology.

Kafka's animal begins by announcing proudly: "I have completed the construction of my burrow and it seems to be successful."[72] The rest of the story calls into question the meaning of "success." Despite his elaborate burrow, the animal worries endlessly about his vulnerability and is constantly preoccupied with defensive measures: "I can scarcely pass an hour in completely tranquility. . . . " His happiest dreams are the ones in which he imagines that he has re-constructed the burrow into a perfectly impregnable fortress. When he awakes, however, he is only too aware of the burrow's defects, only too ready to imagine enemies (perhaps unknown creatures of the inner earth) that seek to invade the burrow and destroy him. He reminds us, though, that he did not build the burrow "simply out of fear." He began to build as a defense against attack, but he found unexpected pleasures in the process of building itself: the joy of labor, the intellectual pleasure of solving sweet technical problems. He takes great pride in his creation: "Here is my castle, which I have wrested from the refractory soil with tooth and claw, with pounding and hammering blows, my castle which can never belong to anyone else. . . . I and the burrow belong so indissolubly together. . . ." But the world he has built becomes the source of anxieties as well as pleasures. When he recalls the "ordinary life" of surface nature, the burrower well remembers its lack of comfort and security. In his life below the surface, his fears "are different from ordinary ones, prouder, richer in content, often long repressed, but in their destruc-tive effects they are perhaps much the same as the anxieties that existence in the outer world gives rise to." Most of "The Burrow" catalogs these "destructive effects." The little creature keeps hearing a whistling sound, which he imagines is some invading enemy. As he tries in all sorts of ways to locate this enemy, he becomes ever more agitated and confused. Finally he realizes that "simply by virtue of being owner of this great vulnerable edifice I am obviously defenseless against any serious attack. The joy of possessing it has spoiled me, the vulnerability of the burrow has made me vulnerable; any wound to it hurts me as if I myself were hit. It is precisely this that I should have foreseen; instead of thinking only of my own defense—and how perfunctorily and vainly I have done even that— I should have thought of the defense of the burrow." For a moment he considers another way of finding security. If another creature should break into the burrow, he muses, perhaps they could come to some sort of understanding—but even as he thinks this he knows very well "that no such thing can happen, and that at the instant when we see each other, more, at the moment when we merely guess at each other's presence, we shall both blindly bare our claws and teeth, . . . both of us filled with a new and different hunger, even if we should already be gorged to bursting. . . . My burrow could not tolerate a neighbor. . . ." He cannot change his habits; for too long he has assumed "I can only trust myself and my burrow." So the burrower tries to keep very quiet, hoping the enemy will go away on its own. "But all remained unchanged"—thus the story ends in mid-page. Kafka is supposed to have finished the tale, ending with the collapse of the burrow and the defeat of the burrower. The conclusion, however, has never been found.[73]

The present book, like Kafka's tale, concludes not so much with an answer as with an anxious foreboding about what might happen if all remains unchanged. In this prolonged thought experiment, we have looked at both subterranean reality and subterranean fiction to ask what might be the psychic and social consequences of human life in a predominantly self-constructed environment. The experiment suggests that the loss of natural surroundings could lead to many other less tangible losses: a diminished sense of reality in surroundings where things seem to be done by magic; the weakening of communal bonds when connections between classes and individuals become subordinate to technological connections; the diminution of personal autonomy when the social environment becomes oppressively authoritarian and when nature ceases to speak as source of independent moral authority; the loss of security as the increasing complexity of the humanmade world and the increasing simplicity of the natural one mesh to promote instability and fragility[74]; and (perhaps most serious of all) the loss of transcendence when maintenance of the material and social structures of the built world becomes the only imaginable end of human life.

For Forster, Villiers, and Tarde—three maverick conservatives—both individual strength and communal strength are found only in relation to some transcendental purpose, a purpose that surpasses the creation and maintenance of the technological environment. These writers demonstrate an inverse relationship between personal prosperity and social bonds. Tarde proclaims that the pursuit of material wealth inevitably divides humanity, whereas the twin ideals of art and love unite it. Forster too sees the quest for individual comfort, which drives people into swaddled isolation, as the enemy of genuine sociability. Villiers shows how technological magic lures us away from living people to impossible, sterile dreams of technical control and perfection. In all cases the technological conquest of nature has its nemesis in the loss of human solidarity. Instead of serving the community, technology becomes an escape from it.[75] For Forster, the essence of community is to be found in the ties of blood, death, and earth; for Villiers, in religious faith; for Tarde, in love and art. They all end their tales with the image of a hero gazing upward at the stars.

We are doing just the opposite. We have reached a point in human history where, like Bulwer-Lytton's coming race or Tarde's neo-Troglodytes, we are preparing to descend below the surface of the earth forever. I mean this in metaphorical terms, but not entirely so. The real surface of our planet is the upper edge of the atmosphere, beyond which lies the frigid and uninhabitable realm of outer space. We have always lived below the surface, beneath the atmospheric ocean, in a closed, sealed, finite environment, where everything is recycled and everything is limited. Until now, we have not felt like underground dwellers because the natural system of the globe has seemed so large in comparison with any systems we might construct. That is changing. What is commonly called environmental consciousness could be described as subterranean consciousness—the awareness that we are in a very real sense not on the earth but inside it.

That awareness can evolve in many directions. Our environment will inevitably become less natural; the question is whether it will also become less human. The human environment is by definition technological to some degree. But if we allow technology to take over our surroundings, they can become inhospitable to human life.

The underground shelter is the emblem of humanity's ancient and honorable quest for truth, power, beauty, and security through technological achievement. As our burrows become more elaborate, however, the needs of the built environment may come to take precedence over the needs of the human builders. Our increased dependence on technological shelter may lead to the weakening of human interdependence, which is another source of security. We should not forget that society too provides shelter, and in many cases a more flexible and effective kind.

Notes

Chapter 1

1. Yi-Fu Tuan, *Man and Nature* (Resource Paper 10, Commission on College Geography, Association of American Geographers, 1971), p. 9. I am indebted to Melvin Kranzberg for urging me to stress the relativity of the concept of a "natural environment" (private communication, December 16, 1987).

The problem of defining what is natural is discussed at length in Tuan's *Topophilia: A Study of Environmental Perception, Attitudes, and Values* (Prentice-Hall, 1974); see especially the diagrams on pp. 104 and 105.

In this study I am using "nature" in what Tuan calls its "demoted sense," as a "'catch-all' term for everything that is not regarded as man-made." As Tuan notes, this interpretation assumes that nature is both wilderness and countryside, "despite the fact that the countryside is, in many instances, a human artifact no less than the city." What are clearly omitted from this concept of nature are "human artifacts and man-made environments." One might wish for a more elegant or coherent concept of nature, but since the "demoted sense" of nature is certainly predominant today it is the appropriate one to use here. See Tuan, *Man and Nature*, pp. 1–2.

2. Philip Shabecoff, "Scientists See Vast Changes in Earth," *New York Times*, November 4, 1987. See also "What Man Has Done to a Hospitable Planet," *New York Times*, November 22, 1987. These articles report on two major symposia dedicated to summarizing the effects of human technologies on the environment: "Man's Role in Changing the Face of the Earth" (Princeton, N.J., June 1955) and "The Earth as Transformed by Human Action" (Worcester, Mass., November 1987). Tuan (*Man and Nature*, p. 9) comments on the 1955 symposium that "the fact of diminishing nature . . . and of human ubiquity is now obvious."

3. Mircea Eliade, *The Forge and the Crucible*, second edition, tr. S. Corrin (University of Chicago Press, 1978 [1956]), pp. 177–178. Recent archaeological research, to be sure, has stressed the cultural advancements that preceded the

agricultural revolution; see William K. Stevens, "Life in the Stone Age," *New York Times*, December 20, 1988.

4. Lisa Belkin, "Skywalks and Tunnels Bring New Life to the Great Indoors," *New York Times*, August 10, 1988; "Atlanta Mall Rules Organ Grinder Obsolete," *New York Times*, October 6, 1988.

5. I am alluding to Sigfried Giedion's classic book *Mechanization Takes Command, a Contribution to Anonymous History* (Oxford University Press, 1948).

6. Eliade, *Forge and Crucible*, p. 181.

7. On this hypothesis of the origins of the pastoral, see Leo Marx, "Literary Culture and the Fatalistic View of Technology," in *The Pilot and the Passenger: Essays on Literature, Technology, and Culture in the United States* (Oxford University Press, 1988), p. 188 (originally published in *Alternative Futures*, 1978).

8. Arthur P. Molella, Mumford in Historiographical Context, draft presented at International Symposium on Lewis Mumford, University of Pennsylvania, Philadelphia, November 1987, pp. 8–9 (revised version forthcoming in *Lewis Mumford: Public Intellectual*, ed. T. P. Hughes and A. Hughes [Oxford University Press]).

9. Lewis Mumford, *Technics and Civilization* (Harcourt, Brace, 1934), p. 77.

10. Ibid., pp. 69–70.

11. Ibid., p. 70.

12. Box 9, section 2, Mumford Collection, Van Pelt Library, University of Pennsylvania.

13. Box 11, folder 2, p. 80, Mumford Collection.

14. The term "environment" is relatively new in English. Thomas Carlyle first used it in 1827, and it passed into common usage in its modern biological sense only in the 1870s. In 1881 the *Fortnightly Review* still felt compelled to define the term for its readers. (See *Oxford English Dictionary*.) In the latter years of the nineteenth century, the French coined the term "la vie factice"—literally, factitious or artificial life. It was popular in literary and artistic circles, but it tended to be restricted to particular man-made objects rather than denoting the environment as a whole. Nowadays the combination of "environment" with the adjective "artificial" has become reasonably common (it provides the title of the first chapter of Theodore Roszak's much-read *Where the Wasteland Ends: Politics and Transcendence in Postindustrial Society* [Doubleday, 1972]). "Artificial" is a somewhat confusing adjective, however. Because it implies a contrast with a state of nature, our understanding of what is artificial keeps shifting along with our understanding of what is natural. Mumford's adjective "manufactured," and the similar one "humanmade," have an advantage over "artificial" in that they are somewhat more objective. They define the environment in terms of its origins rather than in terms of its contrast with the "natural." I use all these terms, as well as the phrase "technological environment," more or less interchangeably in this book.

15. Oswald Spengler, *The Decline of the West*, tr. C. F. Atkinson, volume 2 (Knopf, 1950), p. 94.

16. Quoted by Andrew Lees in "The Metropolis and the Intellectual," in *Metropolis*, ed. A. Sutcliffe (Mansell, 1987); see p. 85.

17. These three books are discussed by Peter Keating in "The Metropolis in Literature," in *Metropolis*, ed. Sutcliffe; see pp. 136–138.

18. As Peter Inch says, urban images have penetrated the subconscious; they are "the decor of modern myths" and have tremendous psychic impact. See "Fantastic Cities," *Architectural Design* 48, no. 2–3 (1978), pp. 117–121.

19. Countless science fiction stories and films are based on the premise of a city so self-enclosed, so remote from nature, so overwhelmingly technological, that it might as well be an underground city. Some examples: Isaac Asimov's *Caves of Steel* (1953); J. G. Ballard's *High-Rise* (1975) and "The Concentration City" (1957); Ridley Scott's *Blade Runner* (1982).

20. Spengler, *Decline*, vol. 2, p. 107.

21. Mumford, *Technics*, p. 159.

22. Lewis Mumford, *Sketches from Life: The Autobiography of Lewis Mumford. The Early Years* (Dial, 1982), p. 56.

23. Lewis Mumford, "Technics and the Nature of Man," *Technology and Culture* 7, no. 3 (1966), p. 303.

24. For the classic versions of these stories, see *Bulfinch's Mythology* (Modern Library, 1934). On Proserpine and Ceres, see pp. 47–52; on Orpheus and Eurydice, pp. 150–153; on Aeneas and the Sibyl, pp. 213–218.

25. See Tuan, *Topophilia*, pp. 129–132, for a general description of the vertical cosmos among preliterate peoples.

26. Ibid., pp. 129, 247.

27. Giles Gunn, *The Culture of Criticism and the Criticism of Culture* (Oxford University Press, 1987), p. 173. As Gunn explains in a footnote, Erich Auerbach and Northrop Frye are the critics who have most thoroughly chronicled this spiritual displacement.

28. William Beckford, *Vathek*, tr. H. B. Grimsditch (London: Folio Society, 1958), p. 114. In the more standard translation by Samuel Henley the first sentence reads: "Their eyes at length growing familiar to the grandeur of the surrounding objects, they extended their view to those at a distance, and discovered rows of columns and arcades, which gradually diminished, till they terminated in a point radiant as the sun when he darts his last beam athwart the ocean." (Quoted in Robert J. Gemmett, *William Beckford* [G. K. Hall, 1977], p. 92.)

29. Quoted in Gemmett, *William Beckford*, p. 93. See also Gemmett's discussion of Beckford's influence: ibid., pp. 137–148.

30. Philip B. Gove, *The Imaginary Voyage in Prose Fiction: A History of its Criticism and a Guide for its Study* (London: Holland Press, 1961 [1941]).

31. The idea of subterranean exploration near the poles was first used in a 1720 story set near Greenland: *La vie, les aventures et le voyage de Groenland du reverend*

Père Cordelier Pierre de Mesange. A selection from the story appears in Pierre Versins, *Outrepart: Anthologie d'Utopies de Voyages extraordinaries et de Science fiction, autrement dit, de Conjectures romanesques rationelles* (Paris: Éditions de la Tête de Feuilles, 1971), pp. 158–162.

32. Mott T. Greene, *Geology in the Nineteenth Century* (Cornell University Press, 1982), pp. 236–257.

33. Henry Faul and Carol Faul, *It Began with a Stone: A History of Geology from the Stone Age to the Age of Plate Tectonics* (Wiley-Interscience, 1983), pp. 101–103.

34. On the lunatic fringe the idea of a hollow earth has never disappeared. In 1906 William Reed published *The Phantom of the Poles* (New York: Walter S. Rockey), and in 1920 Marshall B. Gardner published *A Journey to the Earth's Interior, or, Have the Poles Ever Been Discovered?* (publisher unknown). For a more recent example, see Raymond Bernard, *The Hollow Earth: The Greatest Geographical Discovery in History, Made by Admiral Richard E. Byrd in the Mysterious Land beyond the Poles—The True Origin of the Flying Saucers* (New York: Fieldcrest, 1964), which asserts that Admiral Byrd went 4,000 miles into the polar openings, that "it is very possible that the mysterious flying saucers come from an advanced civilization in the hollow interior of the earth," and that after nuclear war this hollow interior would permit the survival of human life even after fallout destroyed all life on the surface.

Most intriguing of all are the Koreshans, the followers of Cyrus Read Teed of Chicago, who in the late nineteenth century took Symmes's ideas one step further and asserted not only that the earth is habitable within but that we humans are in fact within. The Koreshans believed that the earth is a concave sphere, with all life on its inner surface, and that the sun, an electromagnetic battery, has a light and a dark side and revolves in the center of the universe once every 24 hours. The Koreshans moved to the Gulf Coast of Florida in 1893. In 1897 the Koreshan Unity, as the sect called itself, carried out at Naples, Florida, geodetic measurements using a "rectilineator" that supposedly demonstrated that the curvature of the earth is concave, not convex.

35. Faul and Faul, *Stone*, pp. 101–103; Robert V. Bruce, *The Launching of American Science 1846–1876* (Knopf, 1987), pp. 206–209.

36. Robert Lee Rhea demonstrates Poe's indebtedness to Reynolds's speech in "Some Observations on Poe's Origins," *Texas Studies in English* no. 10 (July 1930), pp. 135–146. Robert F. Almy describes the connection between Reynolds and Poe in "J. N. Reynolds: A Brief Biography with Particular Reference to Poe and Symmes," *Colophon* 2 (Winter 1937), pp. 227–245.

37. "Captain Adam Seaborn" (pseud. John Cleves Symmes), *Symzonia: A Voyage of Discovery* (1820), facsimile reproduction (Gainesville: Scholars' Facsimiles and Reprints, 1965), p. 87. J. O. Bailey describes Poe's indebtedness to Symmes's writings in two articles: "Sources for Poe's *Arthur Gordon Pym*, 'Hans Pfaal,' and Other Pieces," *PMLA* 57 (1942), pp. 513–535; "An Early American Utopian Fiction," *American Literature* 14 (1942), pp. 285–293. Everett F. Bleiler suggests rather the predominant "influence of the common premodern theory that waters of Ocean flowed down through Whirlpools and Chasms into a great

subterranean reservoir (the Abyss of Waters), whence they reemerged as ground water or ocean" (*Science Fiction Writers* [Scribner, 1982], p. 13). He cites the classic exposition of this idea in Athanasius Kircher's *Mundus Subterraneus* (1664).

38. Bailey calls it "the first full-blown American utopia" ("Sources," p. 514n).

39. Quoted in Jean Jules-Verne, *Jules Verne: A Biography,* tr. R. Greaves (Taplinger, 1976 [1973]), p. 60. The essay was published in the April 1864 issue of *Musée des Familles.*

40. Quoted in Peter Costello, *Jules Verne: Inventor of Science Fiction* (Scribner, 1978), p. 82.

41. Lord Kelvin's geological theories are discussed by Stephen Jay Gould in *The Flamingo's Smile: Reflections in Natural History* (Norton, 1985), pp. 126–138.

42. The size of the core could be determined only with the advent of sophisticated seismic techniques. In 1913 Beno Gutenberg made "the first correct interpretation of the arrivals of seismic waves reflected and refracted by the core, and calculated its diameter as about 7000 km, a figure that still stands" (Faul and Faul, p. 216).

For a full and helpful discussion of the relationship between Verne's *Journey* and contemporary geology, see Wendy Lesser, *The Life Below the Ground: A Study of the Subterranean in Literature and History* (Faber and Faber, 1987), especially pp. 41–46.

43. Everett F. Bleiler (*The Guide to Supernatural Fiction* [Kent State University Press, 1983], p. 407) mentions another sequel to *Arthur Gordon Pym:* a novel by Charles R. Dake called *A Strange Discovery.*

44. Wells himself notes these frequent comparisons in his *Autobiography;* see A. J. Hoppe, "Introductory Note," in H. G. Wells, *The Wheels of Chance: The Time Machine* (Dutton, 1935), p. v.

45. Republished in 1910 and 1911 in a revised and altered edition under the title *The Sleeper Awakes.*

46. Tarde actually wrote the *Fragment* in 1884. For information on its composition and publication, see Jean Milet, *Gabriel Tarde et la philosophie de l'histoire* (Paris: Librairie Philosophique J. Brin, 1970), p. 21n.

47. E. M. Forster, introduction to *Collected Short Stories of E. M. Forster* (London: Sidgwick and Jackson, 1947), p. vii.

48. Mark R. Hillegas, *The Future as Nightmare: H. G. Wells and the Anti-Utopians* (Oxford University Press, 1967), pp. 86, 94.

49. H. G. Wells, *The Time Machine and Other Stories* (London: Ernest Benn, 1917), p. 54.

50. Baum introduced the Nome King, Ruggedo, in chapter 2 of *Ozma of Oz* (1907). For the populist Baum, the Nome King was the image of greedy, life-denying capitalism. Ruggedo reappears in *The Emerald City of Oz* (1910) and in *Tik-Tok of Oz* (1914). In *Rinkitink in Oz* (1916) Kaliko becomes the Nome King. In *The Magic of Oz* (1919) the dethroned Ruggedo returns to Oz, which he wants to conquer; he is repulsed.

Baum wrote other books featuring underwater adventures. In *The Sea Fairies* (1911), the little girl Trot and Cap'n Bill end up in the underwater kingdom of mermaids. In *The Scarecrow of Oz* (1915) they are engulfed by a mighty whirlpool and find themselves in a sea cavern; after many perilous adventures there, they return to the Emerald City.

In *Dorothy and the Wizard of Oz* (1908), Dorothy and the Wizard are sucked into the center of the earth and make their way back to Oz through many subterranean perils. In Baum's last Oz book, *Glinda of Oz* (1920), Dorothy, Ozma, and Glinda are imprisoned in a crystal-domed city on an enchanted island just before the island is submerged.

For summaries of plots and characters, see Jack Snow, *Who's Who in Oz* (Reilly & Lee, 1954). See also the bibliography by Michael Patrick Hearn in *The Annotated Wizard of Oz: The Wonderful Wizard of Oz by L. Frank Baum* (Clarkson N. Potter, Inc., 1973).

51. The first Pellucidar story appeared in 1913. In the third and fourth tales of the series (*Tanar of Pellucidar* and *Tarzan at the Earth's Core,* both published in 1929), Tarzan himself enters the inner world, using an enormous zeppelin to sail through a polar opening (evidently Burroughs's version of Symmes's hole).

See Irwin Porges, *Edgar Rice Burroughs: The Man Who Created Tarzan* (Brigham Young University Press, 1975), pp. 162, 469–471, 651, 665, 669. See also Everett Bleiler on Burroughs in *Science Fiction Writers,* ed. E. F. Bleiler (Scribner, 1982), pp. 59–64.

52. See Leo Marx, *The Machine in the Garden: Technology and the Pastoral Ideal in America* (Oxford University Press, 1964), p. 4.

53. In *Literary Theory: An Introduction* (University of Minnesota Press, 1983), Terry Eagleton uses a story about a deep pit as a paradigmatic example of a structuralist interpretation of a text (pp. 95–96).

54. Ibid., p. 109. For an extensive list of stories on subterranean themes see Bleiler, *Guide,* Motif Index, p. 603, under the heading "Subterranean Horrors and Marvels." See also the subheading "Tubes, Subways, and Cars" under "Railroads" (p. 598).

55. The concept of "technological momentum" has been developed primarily by Thomas Parke Hughes, most recently in *Networks of Power: Electrification in Western Society, 1880–1930* (Johns Hopkins University Press, 1983). For a discussion of the concept, see John M. Staudenmaier, *Technology's Storytellers: Reweaving the Human Fabric* (MIT Press, 1985), pp. 148–161.

56. For a fine comparison between scientific theorizing and the making of fiction, see Gillian Beer, *Darwin's Plots: Evolutionary Narrative in Darwin, George Eliot and Nineteenth-Century Fiction* (London: Ark Paperbacks, 1983), especially pp. 89–95 and 159–161.

57. In his preface (written in 1921) to *The Sleeper Wakes* (London: W. Collins Sons, n.d.), p. 5.

58. Clifford Geertz, quoting E. H. Galanter and M. Gerstenhaber in *The Interpretation of Cultures* (Basic Books, 1973), p. 77; cited by Gunn, p. 102.

59. Fredric Jameson, "World-Reduction in Le Guin: The Emergence of Utopian Narrative," in *Science Fiction Studies*, ed. R. D. Mullin and D. Suvin (Gregg, 1976), p. 253. Jameson introduces the concept of world-reduction in analyzing the work of the modern science fiction writer Ursula K. Le Guin, who uses it to imagine a planet without gender (Gethen, in *The Left Hand of Darkness*) or one without animal life other than in human form (Gethen again, or Anarres in *The Dispossessed*). Jameson quotes Le Guin's own description of her invention of a genderless society as a thought experiment: "Einstein shoots a light-ray through a moving elevator; Schrödinger puts a cat in a box. There is no elevator, no cat, no box. The experiment is performed, the question is asked, in the mind."

60. For an excellent discussion of these issues, see Rosemarie Bodenheimer, *The Politics of Story in Victorian Social Fiction* (Cornell University Press, 1988), especially pp. 3–7.

Chapter 2

1. Mumford, *Technics*, p. 70.

2. Quoted by Beer, *Darwin's Plots*, p. 4.

3. Tuan, *Topophilia*, pp. 129, 134, 141, 148, 247. See also Mircea Eliade, *The Myth of the Eternal Return: Cosmos and History* (Princeton University Press, 1954).

4. This is the final line of Hutton's 1788 treatise *Theory of the Earth*, published in the *Transactions of the Royal Society of Edinburgh*. Hutton's book *Theory of the Earth with Proofs and Illustrations* was published in Edinburgh in 1795.

5. Stephen Jay Gould, *Time's Arrow, Time's Cycle: Myth and Metaphor in the Discovery of Geological Time* (Harvard University Press, 1987), pp. 14, 17. Gould's first chapter, "The Discovery of Deep Time," summarizes the issues raised here. Gould credits John McPhee, the modern chronicler of geological science in *Basin and Range* and many other books, with inventing the term "deep time."

6. Ibid., pp. 66–73.

7. Carolyn Merchant, *The Death of Nature: Women, Ecology, and the Scientific Revolution* (Harper & Row, 1980), pp. 1–2.

8. Ibid., pp. 20–30. See also Frank Dawson Adams, *The Birth and Development of the Geological Sciences* (Dover, 1954 [1938]), pp. 286–305.

9. Eliade, *Forge and Crucible*, p. 56.

10. Merchant, *Death*, p. 2.

11. Ibid., pp. 36–38.

12. Carolyn Merchant, "Mining the Earth's Womb," in *Machina Ex Dea: Feminist Perspectives on Technology*, ed. J. Rothschild (Pergamon, 1983), p. 114, quoting Bacon from *The Rise of Modern Science in Relation to Society*, ed. L.

Marsak (Collier-Macmillan, 1964), p. 45. Merchant's article is adapted from chapters 1 and 7 of *The Death of Nature*.

13. Francis Bacon, *Works*, volume 4, ed. J. Spedding, R. Ellis, and D. Heath (Longmans, Green, 1970), pp. 219, 343.

14. Ibid., p. 171. The origins and implications of this reversal of values are discussed in much more detail by Hannah Arendt in *The Human Condition* (University of Chicago Press, 1958).

15. Daniel Worster, *Nature's Economy: A History of Ecological Ideas* (Cambridge University Press, 1977), pp. vii–xii.

16. Before the scientific revolution, organic metaphors served to explain not only nature's significance but also natural events. Geological phenomena were understood as biological functions writ large—the earth fashioned in the image of man. Even in the late seventeenth century the most popular book on the underworld—*Mundus Subterraneus* (1664), by the imaginative and prolific Jesuit scholar Athanasius Kircher—used such an organic analogy to explain the hydrologic cycle. Kircher proposed that numerous holes in the sea bottom open into subterranean passages, so that ocean waters pass into great caverns in the mountains; from there waters flow as springs and rivers back into the sea, passing through the earth's body just as veins and arteries carry blood through the body. (Adams, *Birth and Development* pp. 433–439; see also Roy Porter, *The Making of Geology: Earth Science in Britain 1660–1815* [Cambridge University Press], p. 30.)

17. Porter, *Making of Geology*, p. 35, quoting R. W. T. Gunter, *Early Science in Oxford* (Oxford University Press, 1945), volume 14, p. 270.

18. Porter, p. 57.

19. Ibid., p. 37, quoting from *Philosophical Transactions* of 1666.

20. Porter, pp. 56–57, 60; also Faul and Faul, *Stone*, p. 86.

21. Faul and Faul, p. 86; Porter, pp. 36, 48, 56–58, 60.

22. In the words of Marjorie Hope Nicolson: "There is nothing in the geology of the seventeenth century to correspond to the spectacular discoveries of astronomy. Geology was only part of 'natural Philosophy,' and not an important part. . . . Earth sciences . . . were handicapped by the tendency of generations to read cosmic processes into earth and man, and to find inevitable similarities between the body of earth and the body of man." (Nicolson, "Literary Attitudes Toward Mountains," in *Dictionary of the History of Ideas*, volume 3, ed. P. Wiener [Scribner, 1973], p. 255.) As Nicolson points out, even in the late seventeenth century many geological phenomena were explained by comparing the earth's fall with Adam's fall. In Thomas Burnet's *Sacred Theory of the Earth* (1684), one of the most influential works of the century, the jagged, rugged appearance of the earth is explained as "the Ruines of a broken World," which had been, prior to the Flood, perfectly smooth, regular, and uniform, without "a Wrinkle, Scar or Fracture in all its Body." In his wide travels Burnet visited both mountains and caverns, and observed the effects of earthquakes and volcanoes. For him, all these were examples of the confusion, disorder, and chaos that prevailed upon the fallen earth, now a great ruin of "wild, vast and indigested Heaps of

Stone and Earth" (quoted in ibid., pp. 257–258). Burnet's book was first published in Latin, as *Telluris theoria sacra* (1681). It was again published in Latin in 1689 in an expanded edition, and this edition was republished in English in 1690–91. Gould discusses Burnet's book at length in chapter 2 of *Time's Arrow*.

23. Faul and Faul, pp. 93–94.

24. Adams, pp. 201–204, 211–227. See also Nicolaas A. Rupke, *The Great Chain of History: William Buckland and the English School of Geology (1814–1849)* (Clarendon, 1983), pp. 112–114.

25. Rupke, pp. 111–117.

26. Ibid., p. 130.

27. Ibid., p. 114.

28. William Feaver, *The Art of John Martin* (Clarendon, 1975), p. 146.

29. Rupke, pp. 111–117; Adams, pp. 263–276; Porter, pp. 167–168, 179–180. Because he was so practical and devoted to observation, Smith was a great hero of the Baconians. Many canals were being built to transport coal, and so Smith was often asked to report on coal deposits to determine if building a canal was justified. He also worked on projects to drain bogs and provide irrigation (called to Bath in 1810 when the hot springs failed, he succeeded in restoring their flow). Thus Smith traveled all over England, sometimes as much as 10,000 miles a year, examining geological formations both below and above the surface. He was not particularly knowledgeable about fossils, but he had two good friends (both parsons) who were, and they identified the specimens he unearthed.

30. Rupke, pp. 117–129.

31. Feaver, p. 94; Glyn Daniel, *The Idea of Prehistory* (World, 1962), pp. 41–42; Ruth Moore, *Man, Time, and Fossils: The Story of Evolution* (Knopf, 1953), p. 234.

32. Quoted by Rupke, p. 32. Buckland was well known not only for his scientific work but also for his colorful personality. In the words of Charles Darwin, Buckland, "though very good humored and good natured, seemed to me a vulgar and almost coarse man. He was incited more by a craving for notoriety, which sometimes made him act like a buffoon, than by a love of science." (Quoted by Gould in *Time's Arrow*, p. 99.)

33. Stephen Jay Gould, in "The Freezing of Noah" (*Flamingo's Smile*, pp. 114–125), praises Buckland for his intellectual integrity.

34. As Gould says, one experiences "an almost eerie feeling while reading *Reliquiae diluvianae* in the light of later knowledge about glacial theory. So many of Buckland's specific empirical statements almost cry out for interpretation by ice sheets rather than water. . . . Buckland . . . became one of England's first converts to [Louis Agassiz's] glacial theory." (Ibid., pp. 123–124.) That theory is, of course, still the accepted explanation for most of the evidence Buckland describes so carefully.

35. Moore, p. 234.

36. Quoted by Rupke, p. 91.

37. Arthur Keith, *The Antiquity of Man* (Williams and Norgate, 1915), pp. 46–55, quoted in *The World of the Past,* ed. J. Hawkes (Knopf, 1963), p. 182.

38. Rupke, pp. 94–95.

39. Ibid., p. 90.

40. Ibid., p. 164. Rupke's hypothesis—that only proper scientific skepticism kept Buckland from accepting evidence of ancient human fossils—seems inadequate. In particular, it does not seem sufficient to explain Buckland's tortured reasoning in the case of Kent's Cavern. Rupke discreetly but misleadingly mentions the incident only very briefly (p. 94).

41. Keith, pp. 46–55, quoted in Hawkes, pp. 182–183.

42. Hawkes, pp. 29–30. Boucher exhibited his findings in 1838 in Abbeville and the next year in Paris. In the 1840s he continued to accumulate evidence from other local excavations: ones made to repair the fortifications of Abbeville, to dig flint for making roads, to gather loam for making bricks, and to lay the foundation for a hospital. He found bones and teeth of extinct elephants, rhinoceroses, bears, and other mammals mixed together with flint tools and stone axes.

43. In Lyell's words, Pengelly (who directed the excavation) "had the kindness to conduct me through the subterranean galleries after they had been cleared out in 1859; and I saw, in company with Dr. Falconer, the numerous fossils which had been taken from the subterranean fissures and tunnels, all labelled and numbered, with references to a journal kept during the progress of the work, and in which the geological position of every specimen was recorded with scrupulous care" (Lyell, pp. 93–105, in Hawkes, p. 153).

44. Lyell, pp. 75–79, in Hawkes, p. 168.

45. Quoted by Moore, p. 235.

46. Quoted by Beer, *Darwin's Plots,* pp. 138–139.

47. Hawkes, p. 78. The first cave art was discovered in a Spanish cave in the summer of 1879. A little girl, whose father was interested in prehistory and was excavating there, wandered off into the cave and ran back to her father exclaiming, "Papa, mira toros pintados!" Her father was unable to bring others around to his conviction that these paintings were the work of paleolithic cavedwellers, and was accused of foolishness or even fraud.

48. Lubbock was the first to put these terms into wide circulation. They had been introduced in 1851 when Daniel Wilson had published *The Archaeology and Prehistoric Annals of Scotland.* See Daniel, *Idea,* p. 13; also Glyn Daniel, *The First Civilizations: The Archaeology of Their Origins* (Thames and Hudson, 1968), pp. 16–17, 192.

49. Quoted by Daniel, *Idea,* p. 63.

50. C. W. Ceram [Kurt W. Marek], *The March of Archaeology,* tr. R. and C. Winston (Knopf, 1953), p. 311.

51. Quoted by Hawkes, p. 49. Belzoni grabbed as many items as he could, intending to send or sell them to British museums. At one point agents of the

French consul held him up at gunpoint when he was trying to deliver to a British consul an obelisk the French claimed was theirs (ibid.). Other anecdotes about Belzoni and other early archaeologist-explorers are recounted in Bruce Norman's *Footsteps: Nine Archaeological Journeys of Romance and Discovery* (Salem House, 1988).

52. Both were wrong, but this became clear only after Henry Rawlinson deciphered cuneiform writing in 1846.

53. Seton Lloyd, *Mesopotamia,* quoting Wallis Budge; in Hawkes, p. 401.

54. Daniel, *Idea,* p. 184. To see how archaeological discoveries were reported in the pages of *The Illustrated London News* beginning in 1842, consult *The Great Archaeologists,* ed. E. Bacon (Bobbs-Merrill, 1976).

55. Lloyd, pp. 1–26, in Hawkes, p. 403.

56. Hawkes, p. 59; see also J. Forde-Johnston, *History from the Earth: An Introduction to Archaeology* (Phaidon, 1974), p. 58. For a discussion of the relationship between poetry and archaeology, with particular reference to Schliemann, see Lesser, *Life Below the Ground,* pp. 69–76.

57. American excavators in the 1930s meticulously divided the site into 46 building periods (Hawkes, p. 59). The tells of the Middle East are man-made artifacts that resemble geological forms; the causes of their degradation are both cultural and natural, and since the 1870s archaeologists have understood that excavating them requires geological as well as archaeological understanding. For a recent summary of geological contributions to archaeological investigations, see Arlene Miller Rosen, *Cities of Clay: The Geoarcheology of Tells* (University of Chicago Press, 1986).

58. Daniel, *Idea,* 69.

59. Hawkes, p. 52.

60. Forde-Johnston, p. 56.

61. Layard, quoted in Hawkes, p. 296.

62. Carter, in Hawkes, pp. 500–501. Carter published the account of his discovery of Tutankhamen's tomb in three volumes between 1922 and 1933. The account was republished in one volume as *The Tomb of Tutankhamen* (Dutton, 1972 [1954]).

63. Ibid., p. 501.

64. Francis D. Klingender, *Art and the Industrial Revolution,* rev. and ed. Arthur Elton (Schocken, 1970 [1947]), pp. 120–121.

65. Paul Bourget, "Discours académique de 13 juin 1895. Succession à Maxime Du Camp," in *L'anthologie de l'Académie française,* volume 2 (Paris, 1921), p. 191ff; quoted by Walter Benjamin in *Charles Baudelaire: A Lyric Poet in the Era of High Capitalism,* tr. Harry Zohn (NLB, 1973), p. 86.

66. Arthur Conan Doyle, *The Maracot Deep and other Stories* (Doubleday, 1929 [1927]), p. 18.

67. Ibid., p. 60.

68. On the subject of Bulwer-Lytton's influence on the revival of supernatural fiction in the 1840s and on the occult novel in the 1890s, see James L. Campbell, Sr., *Edward Bulwer-Lytton* (Twayne, 1986), pp. 127, 133.

69. Doyle, pp. 186, 188. Doyle's interest in the possibility that the earth is a living, organic being is evident in "When the World Screamed," his story about a "mad scientist" who drills down to pierce the earth's "skin."

70. Victor Hugo, *Les Misérables,* rev. and ed. L. Fahnestock and N. MacAfee (New American Library, 1987), p. 983.

71. Karl Marx, *Capital: A Critique of Political Economy* (London, 1974), volume 1, pp. 279–280; quoted by Donald MacKenzie, "Marx and the Machine," *Technology and Culture* 25, no. 3 (1984), p. 480.

72. These debates are discussed in a clear and helpful way in MacKenzie's article (ibid.). For a discussion with particular relevance to literature, see Terry Eagleton, *Marxism and Literary Criticism* (University of California Press, 1976), pp. 3–6.

73. Sigmund Freud, *The Standard Edition of the Complete Psychological Works* (Hogarth, 1950), volume 18, pp. 27–29; quoted and discussed by Wolfgang Schivelbusch in *The Railway Journey: The Industrialization of Time and Space in the 19th Century* (University of California Press, 1986 [1977]), pp. 164–166.

74. Hans Aarsleff, "Taine and Saussure," in *From Locke to Saussure: Essays on the Study of Language and Intellectual History* (University of Minnesota Press, 1982), pp. 356–371. I want to thank William Johnston, professor of history at the University of Massachusetts at Amherst, for calling this article to my attention.

75. Peter Caws, "Structuralism," in *Dictionary of the History of Ideas,* volume 4, pp. 326–329.

76. See Terry Eagleton, *Literary Theory: An Introduction* (University of Minnesota Press, 1983), pp. 91–126.

77. Gunn, *Culture of Criticism,* pp. 48–49.

78. Jacques Derrida, *Of Grammatology,* tr. Gayatri Spivak (Johns Hopkins University Press, 1976), p. 14; quoted in Gunn, pp. 49–50.

79. Eagleton, *Literary Theory,* pp. 137, 139, 145.

80. Having resolved to focus on British and French writers, I am reluctantly omitting a treatment of Henry David Thoreau's complex discussion of the metaphorical as opposed to the factual dimensions of Walden Pond's depth. In the chapter "The Pond in Winter," Thoreau (in true Baconian spirit) decides to measure the depth of the pond for himself, having heard stories that Walden had no bottom "or even reached quite through to the other side of the globe." In the same paragraph in which he gives the scientific results, however, Thoreau pivots to poetic language; he is "thankful that this pond was made deep and pure for a symbol." For a full and perceptive discussion, see Walter Benn Michaels, "*Walden*'s False Bottoms," in *Glyph I* (Johns Hopkins University Press, 1977). Toward the end of *Walden,* Thoreau urges his readers to search for truth not horizontally but vertically: ". . . it is easier to sail many thousand miles

through cold and storm and cannibals, in a government ship [he is referring here to the Wilkes expedition] . . . than it is to explore the private sea, the Atlantic and Pacific Ocean of one's being alone. . . . Yet do this even till you can do better, and you may perhaps find some 'Symmes' Hole' by which to get at the inside at last." See also Thoreau's passage on the deep railroad cut near Walden, which he describes as a sort of man-made underworld ("the whole cut impressed me as if it were a cave with its stalactites laid open to the light"), and Leo Marx's discussion of this passage (*Machine*, p. 261) and his general discussion of Thoreau ("Henry Thoreau: Excursions," in *Pilot and Passenger*).

Chapter 3

1. Mumford, *Technics*, p. 74.

2. The Latin word *infra*, meaning "down" or "beneath" or "under," has long been used in English as a prefix denoting the opposite of *super* or *supra*, but mainly in medical contexts. *Infrastructure* was coined in this century.

3. Sigfried Giedion, *Space, Time and Architecture* (Harvard University Press, 1978), p. 739.

4. Box 8, folder 6, chapter 3, "Cities," p. 25, Lewis Mumford Collection, Van Pelt Library, University of Pennsylvania. Mumford doubtless owes some of his consciousness of the relation between the surface and subsurface cities to his mentor, the Scottish city planner, biologist, and social thinker Patrick Geddes. Geddes's biographer Philip Boardman tells us that when Geddes arrived in the United States in 1923 he remarked, upon glimpsing the Manhattan skyline, "I think the skyline is even more strikingly beautiful than in 1900—from a distance, of course. But I'll wager that down underneath New York is still the same: an excellent working-model of hell!" (Philip Boardman, *Patrick Geddes: Maker of the Future* [University of North Carolina, 1944], pp. 398–399.) In *Technics* Mumford still hoped that "with a more carefully planned relationship between working and living . . . the network of tunnels and subways that make up a modern metropolis will become obsolete" (box 8, folder 4, Van Pelt Library).

5. Joel A. Tarr and Gabriel Dupuy, *Technology and the Rise of the Networked City in Europe and America* (Temple University Press, 1988).

6. Klingender, *Art in the Industrial Revolution*, p. 178. See also Kenneth Clark, *Civilisation: A Personal View* (Harper & Row, 1969), pp. 337–339.

7. Alan Trachtenberg, "Foreword," in Schivelbusch, *Railway Journey*, p. xv.

8. Karoly Szechy, *The Art of Tunnelling*, second English edition (Budapest: Akademiai Kiado, 1973 [1966]), pp. 37–38; Cyril C. Means, Jr., "Ancient Tunnels," in F. P. Davidson (ed.), *Tunneling and Underground Transport: Future Developments in Technology, Economics, and Policy* (Elsevier, 1987), pp. 7–9.

9. Mumford, *Technics*, p. 67.

10. Daniel Defoe, *A Tour Thro' the Whole Island of Great Britain*, first edition, 1724–1727; repr. two volumes (London, 1968), volume 2, p. 659, quoted in Schivelbusch, p. 2.

11. J. A. S. Ritson, "Metal and Coal Mining," in *A History of Technology*, ed. C. Singer et al. (Oxford University Press, 1958), volume 4, p. 79.

12. "In 1778 more than 70 Newcomen engines were working in Cornwall, but by 1790 all but one of them had disappeared, having been replaced by the much more economical Boulton and Watt engine." Ibid., p. 78.

13. Schivelbusch, pp. 2–4; Patrick Beaver, *A History of Tunnels* (Citadel Press, 1973 [1972]), p. 55; C. N. Bromhead, "Mining and Quarrying to the Seventeenth Century," in Singer et al., volume 2, p. 1.

14. William H. McNeill, "The Eccentricity of Wheels, or Eurasian Transportation in Historical Perspective," *American Historical Review* 92, no. 5 (December 1987), p. 1111.

15. A. W. Skempton, "Canals and River Navigation before 1750," in *A History of Technology*, ed. Singer et al., volume 3, p. 466.

16. Klingender, pp. 14–16; Beaver, p. 31; Richard Shelton Kirby and Philip Gustave Laurson, *The Early Years of Modern Civil Engineering* (Yale University Press, 1932), p. 168. For an overview of the topic in Britain, see *Transport in the Industrial Revolution*, ed. D. H. Aldcroft and M. J. Freeman (Manchester University Press, 1983).

17. Schivelbusch, p. 19.

18. In addition, English rolling stock used rigid axles that would derail on sharp curves. American rolling stock, by the early 1840s, was designed to negotiate curves much more readily. Ibid., pp. 96–99.

19. Hans Straub, *A History of Civil Engineering*, tr. Erwin Rockwell (MIT Press, 1964), pp. 168–169. See also Sir Harold Harding, "Tunnels," in *The Works of Isambard Kingdom Brunel: An Engineering Appreciation*, ed. A. Pugsley (Cambridge University Press, 1976), pp. 25–50.

20. Beaver, p. 62.

21. For a recent, detailed study of tunneling technology from the 1830s to the present (both hard- and soft-rock techniques), see Graham West, *Innovation and the Rise of the Tunnelling Industry* (Cambridge University Press, 1988).

For a summary of recent advances in tunneling technology, see Walter Sullivan, "Progress in Technology Revives Interest in Great Tunnels," *New York Times,* June 24, 1986. The longest passenger tunnel in the world at the present is the Seikan Tunnel (actually three tunnels, 34 miles in all, of which 14½ miles are underwater) connecting the islands of Honshu and Hokkaido in Japan. The average advance in boring the Seikan Tunnel was 100 feet per day, whereas on the Simplon Tunnel progress averaged 96 feet per week. Sullivan points out that, despite the much-improved knowledge of geological structures and the use of new technological devices (especially giant mechanical "moles"), tunneling remains a perilous enterprise; four serious floods were encountered in digging the Seikan Tunnel.

The tunnel now being dug under the English Channel will be 31 miles long and will include three bores: two railroad tunnels and a central service tunnel. Nine million cubic yards of spoil will be excavated during this project, which is scheduled to be completed and ready for service in 1993. (William G. Miller, "Chunnel builders say 3d time's a snap," *Boston Globe*, December 7, 1987.)

22. Clark, *Civilisation*, p. 330.

23. Klingender, p. 158. For a discussion of Bury, Bourne, and other railway artists, see pp. 150–163.

24. Bodenheimer, *Politics*, p. 148.

25. Samuel Smiles, *Lives of the Engineers* (1861–62), volume 3, pp. 321–323, quoted by Klingender, p. 172.

26. Ibid.

27. Kirby and Laurson, p. 168.

28. William B. Meyer, "The Long Agony of the Great Bore," *American Heritage of Invention and Technology* 1, no. 2 (fall 1985), pp. 52–57. For rates of advance, see table from "A Century of Tunnelling," the Thomas Hawksley Lecture by W. T. Halcrow, delivered to the Institution of Mechanical Engineers in 1941, reproduced in Rolt Hammond, "Historical Development," in *Tunnels and Tunnelling*, ed. C. A. Pequignot (London: Hutchinson, 1963), p. 4. Beaver (p. 67) says the average monthly rate of advance increased from 47 to 126 feet after nitroglycerine began to be used, but he also points out that the project was put under new and better management early in 1869.

29. In the 1930s, when the Union Carbide Corporation bored Hawk's Nest Tunnel through Gauley Mountain in West Virginia to transport water from a river to a power plant, as many as 700 men died and many more were disabled by lung diseases. Three-quarters of these workers were poor blacks who, if they refused to enter the deadly tunnel head, were sometimes clubbed by foremen. According to a recent epidemiological study, the drilling of the Hawk's Nest Tunnel rates as "America's worst industrial disaster." This is the subtitle of Martin Cherniack's book *The Hawk's Nest Incident* (Yale University Press, 1986).

30. Schivelbusch, p. 23.

31. Ibid., p. 30.

32. Ibid., p. 131.

33. Charles Dickens, *Dombey and Son* (New York: John B. Alden, 1883), pp. 66–67.

34. Eagleton, *Marxism and Literary Criticism*, p. 36.

35. Dickens, *Dombey and Son*, p. 214.

36. Ibid. For a discussion of Dickens's response to "progress" in *Dombey* and in other works, see Herbert Sussman, *Victorians and the Machine: The Literary Response to Technology* (Harvard University Press, 1968), pp. 41–75.

37. Richard Ayton, *A Voyage round Great Britain undertaken in the Summer of 1813* (1814), volume 2, pp. 155–160, quoted in *"Hard Times": Human Documents of the Industrial Revolution*, ed. E. Royston Pike (Praeger, 1966), pp. 252–253.

38. Klingender, p. 126; cf. pp. 121–125 on Martin and pp. 127–133 on others who used images of hell to describe industry. See also Sussman, pp. 50–51.

39. Feaver, *Art of John Martin*, pp. 78–79. Martin has grown up among the coal pits and lead mines of Newcastle, where as a boy he had heard many stories of pit disasters. He maintained a lifelong interest in improving mine safety, and he often lamented that he had not been an engineer (see ibid., pp. 113–187). His compelling drawings of hell reached multitudes; he was one of the most famous artists of the day, and his illustrations for *Paradise Lost* were widely reproduced and admired.

40. Klingender, p. 126.

41. John Britton, *Autobiography* (1850), volume 1, pp. 128–129 (large paper edition), quoted in Klingender, p. 104. I want to thank Leo Marx for suggesting (in private conversation) that I think of subterranean imagery in terms of concentric circles expanding from the central image of the mine.

42. Dickens, *The Old Curiosity Shop*, chapter 45, quoted in Klingender, p. 130.

43. See the examples collected by Humphrey Jennings in *Pandaemonium 1660–1886: The Coming of the Machine as Seen by Contemporary Observers*, ed. M.-L. Jennings and C. Madge (Free Press, 1985), especially in the "Theme Sequence" titled "Daemons at Work" (pp. 361–362). See also the listing titled "Miners" (p. 362).

44. George Orwell, *The Road to Wigan Pier* (Penguin, 1963 [1937]), pp. 15–16.

45. Ibid., p. 17.

46. Ibid., pp. 104–105. In chapter 4 Orwell makes it clear that most workers of the 1930s did not enjoy the domestic comforts he is describing; the ideal picture of after-dinner relaxation, he says, is drawn from a well-off working-class family in the prewar years, when England was still prosperous. Orwell explains that for him this scene of working-class prosperity represents a middle ideal between the raw nature of the middle ages (a windowless hut, a chimneyless wood fire, mouldy bread, dirt and disease and early death) and the artificial environment of the future (invisible heaters, furniture made from steel and glass, no dogs or horses and few children).

47. Hugo, pp. 991–994.

48. Schivelbusch, pp. 181–182.

49. David H. Pinkney, *Napoleon III and the Rebuilding of Paris* (Princeton University Press, 1956), p. 132. (See chapter 6, "Paris Underground.") The best primary source is Eugène Belgrand, *Les Travaux souterrains de Paris* (Paris, 1873–1877).

50. Pinkney, p. 143.

51. Henry Law, George R. Burnell, and Daniel Kinnear Clark, *The Rudiments of Civil Engineering*, sixth edition (London: Crosby Lockwood and Co., 1881), p. 586.

52. Szechy, pp. 41, 43.

53. Jules Romains, "L'Équipe du Métropolitain," *Puissances de Paris,* seventh edition (Paris: Éditions de la Nouvelle Revue Française, 1919), pp. 129–131 (my translation).

54. W. M. Patton, *A Treatise on Civil Engineering,* first edition (Wiley, 1895), p. 549. See pp. 544–549 on excavation techniques in general.

55. Klingender, p. 211. The tunnel never made as much money as hoped, however, and it was converted from pedestrian to railway use in 1869. It is now part of the Metropolitan Railway.

56. Beaver (p. 72) claims an even shorter period: 6 months.

57. There were even plans for a Dover-Calais tunnel. Advance headings were dug in 1881–82. Over 2,000 yards were cut on each side before the British fear of invasion from France put a halt to the project.

58. Pequignot, pp. 300–303, 315, 325.

59. W. A. Starrett, *Skyscrapers and the Men Who Build Them* (Scribner, 1928), p. 29.

60. Carl W. Condit, *The Rise of the Skyscraper* (University of Chicago Press, 1952), p. 199. The problem of building foundations in Chicago was aggravated when, beginning in 1904, the Chicago Tunnel Company began to construct small freight tunnels along the main streets of the Loop. These tunnels (9 feet high, 6 feet wide, and 50 feet below the ground surface) caused many structures to settle and created the need for widespread underpinning. See Ralph B. Peck, *History of Building Foundations in Chicago,* University of Illinois Engineering Experiment Station Bulletin Series no. 373, volume 45, no. 29 (January 2, 1948), passim. See also Starrett, pp. 142–145.

61. David McCullough, *The Great Bridge* (Simon and Schuster, 1972), pp. 174, 220, 246–247. See also the description of more typical pneumatic caissons in chapter 13 of Starrett.

62. H. Shirley Smith, "Bridges and Tunnels," in *A History of Technology,* ed. Singer et al., volume 5, pp. 514–516, 520.

63. Benson Bobrick, *Labyrinths of Iron* (Newsweek Books, 1981), p. 151.

64. Jules Romains, "The Sixth of October," volume 1 in *Men of Good Will,* tr. Warre B. Wells (Knopf, 1933), p. 7.

65. Mumford, *Technics,* pp. 245–246.

Chapter 4

1. Mumford, *Technics,* p. 70. On Mumford's principle of organicism, as it applies to art and to society, see Leo Marx, "Lewis Mumford: Prophet of Organicism," presented to the International Symposium on Lewis Mumford, University of Pennsylvania, Philadelphia, 1987, second draft, pp. 9–13 (forthcoming in *Lewis Mumford: Public Intellectual,* ed. Hughes and Hughes).

2. Another modern story of an underworld populated by reptilian horrors is one by A. Merritt (regarded by many as his best) called "The People of the Pit," first printed in the *All-Story Weekly* of January 5, 1918, and reprinted numerous times (e.g., *Amazing Stories*, March 1927; *Amazing Stories Annual*, 1927; *Fantastic*, March 1966; Sam Moskowitz (ed.), *Masterpieces of Science Fiction* [World, 1966]).

3. Jerome Stolnitz, "Ugliness," in Paul Edwards (ed.), *Encyclopedia of Philosophy* (Macmillan and Free Press, 1967), vol. 8, pp. 174–176.

4. The phrase is used by Jerome Stolnitz in "Beauty," *Encyclopedia of Philosophy*, vol. 1, p. 264.

5. Walter John Hipple, Jr., defines the central quest of eighteenth-century aesthetics as "the specification and discrimination of certain kinds of feelings, the determination of the mental powers and susceptibilities which yielded those feelings, and of the impressions and ideas which excited them." (*The Beautiful, The Sublime, and The Picturesque in Eighteenth-Century British Aesthetic Theory* [Southern Illinois University Press, 1957], p. 305.)

6. Ibid., p. 265.

7. Samuel H. Monk, *The Sublime: A Study of Critical Theories in XVII-Century England* (Modern Language Association of America, 1935), pp. 10–24. For a more detailed study of the term, see Theodore E. B. Wood, *The Word "Sublime" and its Context 1650–1760* (Mouton, 1972).

8. "The Pleasures of the Imagination" is the title of Addison's essay published in *The Spectator* in 1712. See Hipple, pp. 13–16.

9. Marjorie Hope Nicolson, "Sublime in External Nature," vol. 4, *Dictionary of the History of Ideas*, p. 335.

10. Ibid.

11. Hipple remarks that everyone after Burke "either imitates him or borrows from him or feels it necessary to refute him" (p. 83). In the 1740s three long excursion poems were published, all of which were widely read and admired: James Thomson's *Seasons*, Mark Akenside's *The Pleasures of the Imagination*, and Edward Young's *Night Thoughts*. These poems obviously influenced Burke. (Nicholson, "Sublime," *DHI*, pp. 335–336.)

12. Edmund Burke, *Sublime and Beautiful*, second edition, in *The Works of the Right Honourable Edmund Burke* (Oxford University Press, 1906), vol. 1, pp. 91–92, 102; quoted in Hipple, p. 88.

13. Quoted by Monk, p. 92.

14. Nicolson, "Sublime," *DHI*, p. 337.

15. Thomas Wallace Knox, *The Underground World* (J. B. Burn, 1880), p. 213.

16. Feaver, *Art of John Martin*, p. 89. For general information on early cave tourism see Porter, *Making of Geology*, p. 102, and Knox, p. 457. For a poem describing a mine tour taken by two English ladies in 1755, see Klingender, *Art in the Industrial Revolution*, pp. 21–22. Other general information on eighteenth-century cave tourism was provided by Emily Davis Mobley, owner of Speleobooks in Schoharie, New York, in a private conversation, March 16, 1988.

17. Marjorie Hope Nicolson, *Mountain Gloom and Mountain Glory* (Cornell University Press, 1959). See also Nicolson, "Literary Attitudes Toward Mountains," *DHI*, volume 3, pp. 253–260; Tuan, *Topophilia*, pp. 70–74.

18. Porter, p. 102. Also see Monk, pp. 203–232.

19. Hamilton, "Some particulars of the present state of Mount Vesuvius," *Philosophical Transactions* 76, pp. 365–380, quoted in Porter, p. 161 (see also p. 124).

20. Quoted in Christopher Thacker, *The Wildness Pleases: The Origins of Romanticism* (St. Martin's, 1983), p. 148; see also pp. 131 and 145–147.

21. Marcia Pointon, "Geology and landscape painting in nineteenth-century England," in *Images of the Earth: Essays in the History of the Environmental Sciences,* ed. L. J. Jordanova and R. S. Porter (British Society for the History of Science, 1979), p. 86.

22. In an 1859 review of John Brett's painting *Val d'Aosta;* quoted by Pointon (ibid.), p. 93.

23. Pointon, pp. 151–152. Wright's painting of Vesuvius closely resembles a painting by John Martin, done just before he died in 1854, titled "The Great Day of his Wrath." Klingender writes: "According to his son Leopold, it was inspired by a journey through the Black Country in the dead of night: 'The glow of the furnaces, the red blaze of light, together with the liquid fire, seemed to his mind truly sublime and awful. He could not imagine anything more terrible even in the regions of everlasting punishment. All he had done or attempted in ideal painting fell far short, very far short, of the fearful sublimity.'" (Klingender, p. 132)

24. This ideological dimension of sublimity is well appreciated by Leo Marx, who pointedly uses the term "the rhetoric of the technological sublime" in showing how sublime images of great and titanic powers were used by publicists and politicians to support a progressive ideology. The concept of technological sublimity was first articulated by Marx in *The Machine in the Garden* (see especially pp. 195–207, 214, 230, and 294–295). Marx drew on Perry Miller's discussion of technology and sublimity in *The Life of the Mind in America.* Miller discusses technology as an element of the sublime primarily in book 3, chapter 1 ("Technological America"), section 4 ("Utility within Universality") in his unfinished opus *The Life of the Mind in America: From the Revolution to the Civil War* (Harcourt, Brace & World, 1965). Miller too focuses on the rhetoric of technological sublimity. He asks (rhetorically), "Could this technological majesty join with the starry heavens above and the moral law within to form a peculiarly American trinity of the Sublime?" The answer is Yes, as Miller shows how American rhetoricians accomplished a "happy reconciliation of the mind to applicability" (p. 291).

Other cultural historians have reiterated the importance of sublimity in the expression of responses to industrialization. See, for example, Sussman, *Victorians and the Machine,* esp. pp. 28–31 (on Carlyle), 45 (on Dickens), and 229; Michael Smith, *Pacific Visions: California Scientists and the Environment 1850–1915* (Yale University Press, 1988), pp. 73–89 passim, 95–96, 102; John Kasson, *Civilizing the Machine: Technology and Republican Values in America, 1776—1900*

(Grossman [Viking], 1976), pp. 161–180. Klingender and Jennings also discuss in general terms the way industrial machines and installations were described as eruptions from the underworld (in Jennings, see especially the selections listed under the theme sequence "Daemons at Work" on pp. 361–362). See also Raymond Williams, "The Welsh Industrial Novel," in *Problems in Materialism and Culture: Selected Essays* (Verso, 1980), pp. 213–214.

25. Eagleton, *Literary Theory*, pp. 206–207.

26. From *The Funeral Sermon of the Felling Colliery Sufferers*, by the Rev. John Hodgson, published in Newcastle in 1813; quoted in Jennings, pp. 132–133.

27. Ibid.

28. Ibid., p. 134.

29. For this reason, William Leiss has connected the technological sublime to the concept of degeneration: the machine is alive and masterful, humanity is degenerate and even moribund. ("Technology and Degeneration: The Sublime Machine," in *Degeneration: The Dark Side of Progress*, ed. J. E. Chamberlin and S. L. Gilman [Columbia University Press, 1985], pp. 145–164.)

30. Monk, p. 97.

31. Sigfried Giedion, *Architecture and the Phenomena of Transition: The Three Space Conceptions in Architecture* (Harvard University Press, 1971), pp. 146, 148. According to Giedion, the first space conception, in which architecture is conceived in terms of space-radiating volumes, emerged with the first high civilizations in Egypt and Mesopotamia and lasted through Greece. A third space conception, synthesizing volume and space (so that exterior and interior space interpenetrate), seems to be emerging in the twentieth century. Although Giedion describes the first and third space conceptions in the book, he deals mainly with the second space conception, which is Roman in origin.

32. Ibid.

33. Ibid., pp. 153–159. According to Giedion's analysis, Gothic architecture is the prime exception to the emphasis on creating interior space; he treats the Middle Ages as a long interlude after which the West again turned to Rome and relearned the techniques of working with large internal spaces. Gothic architecture stresses verticality, as well as the interpenetration of exterior and interior, rather than confined space (pp. 253–256).

34. Ibid., pp. 148, 152, 214.

35. Ibid., p. 160.

36. Piranesi was an archaeologist at a time when the word was rarely used. He literally dug up the foundations of Roman buildings, and he camped for long periods at the more distant sites. He took great pains to figure out how the buildings were constructed, and put numbers and notes on his plates to convey this information to the viewer. He also sold antiquities on the side. In my discussion I shall rely primarily upon Marguerite Yourcenar's fine essay titled "The Dark Brain of Piranesi" (a line from one of Victor Hugo's poems). (M. Yourcenar, *The Dark Brain of Piranesi and Other Essays*, tr. Richard Howard in

collaboration with the author [Farrar, Straus, and Giroux, 1984].) On Piranesi as an archeologist, see pp. 91–99.

37. The actual title of these engravings is *Invenzioni Caprice di Carceri*.

38. Yourcenar, p. 111.

39. Ibid., pp. 104, 111. Yourcenar also points out that the second series of the "imaginary prisons" engravings are noticeably blacker than the earlier impressions but that this darkening may be due to the need to reengrave the cross-hatching (which tended to blur) rather than to psychological reasons (p. 128).

40. Ibid., p. 111.

41. Ibid., pp. 126. Yourcenar quotes Thomas de Quincey's description of Coleridge's recollection of Piranesi's engravings.

42. J. O. Bailey, "The Geography of Poe's 'Dream-Land' and 'Ulalume,'" *Studies in Philosophy* 45 (1948): 512–523.

43. Translated by Wallace Fowlie in Charles Baudelaire, *Les Fleurs du mal et oeuvres choisies,* ed. W. Fowlie (Bantam, 1964), pp. 79–81.

44. See Herbert Dieckmann, "Theories of Beauty to the Mid-Nineteenth Century," *DHI,* vol. 1, pp. 195–206.

45. From a letter of Fanny Kemble to a friend, printed in her *Records of a Girlhood* (1878) and quoted in Jennings, pp. 168–169.

46. For descriptions of the literary and artistic sources of *Pompeii,* see Curtis Dahl, "Recreations of Pompeii," *Archaeology* 9 (1956): 182–191, and "Bulwer-Lytton and The School of Catastrophe," *Philosophical Quarterly* 32 (1953): 428–442. Bulwer-Lytton had the good fortune to publish the book in a year when Vesuvius again erupted (James C. Simmons, "Bulwer and Vesuvius: The Topicality of *The Last Days of Pompeii*," *Nineteenth-Century Fiction* 24 [1969]: 103–105).

47. Edward Bulwer-Lytton, *Vril: The Power of the Coming Race* (Rudolf Steiner Publications, 1972), pp. 14–20, passim. The idea of an underground race had come to Bulwer-Lytton many years earlier. In 1835 he had published a satiric piece titled "Asmodeus at Large" in a collection called *The Student.* The narrator, who suffers from ennui, lives for a time in the underground city of Cyprolis. He meets a wizard, travels to the center of the earth, and sees a gigantic stone figure with countless strings emerging from it. See *Science Fiction Writers,* ed. Bleiler, p. 87; Campbell, *Edward Bulwer-Lytton,* p. 126.

48. Bulwer-Lytton, p. 32.

49. Ibid., p. 126.

50. Ibid., p. 45.

51. Jules Verne, *Voyage au centre de la terre* (Paris: Hetzel, 1864), p. 139; quoted in Michel Durr, "Jules Verne et l'électricité," in *La France des Électriciens 1880–1980,* ed. F. Cardot (Presses Universitaires de France, 1985), p. 338 [my translation].

52. Jules Verne, *The Mysterious Island* (Grosset & Dunlap, n.d.), pp. 457–458.

53. Jules Verne, *The Underground City, or the Black Indies* (Vincent Parke, 1911), pp. 337–338.

54. Jules Verne, *Twenty Thousand Leagues under the Sea* (Scribner, 1933), p. 72.

55. Ibid., pp. 75–81, passim.

56. Durr, p. 341; my translation. Verne, always concerned with technological plausibility, explains that Nemo has used "Bunsen's contrivances, not Ruhmkorff's," and that the electrical motors turn the ship's propellers at 120 revolutions per second. In fact, Verne confuses Bunsen's batteries with Ruhmkorff's coils, and the propeller speed he gives is impossibly high.

57. Raymond Williams, "Utopia and Science Fiction," in *Problems*, p. 201.

58. See Jean Chesneaux, *Jules Verne: Une lecture politique* (François Maspero, 1982 [1971]), especially chapters 2 and 4. This book is available in English as *The Political and Social Ideas of Jules Verne*, tr. Thomas Wikeley (Thames and Hudson, 1972).

59. William Delisle Hay, *Three Hundred Years Hence; or, a Voice from Posterity* (London: Newman and Co., 1881), pp. 176–177.

60. Ibid., pp. 179–180.

61. Ibid., pp. 180, 182.

62. Ibid., p. 230.

63. Ibid., p. 256.

64. Gabriel Tarde, *Underground Man*, tr. Cloudesley Brereton, intro. H. G. Wells (Hyperion, 1974 [reprint of 1905 Duckworth edition]), p. 44.

65. Ibid., p. 57.

66. Ibid., pp. 96–97.

67. McCullough, *Great Bridge*, pp. 197–198.

68. William T. O'Dea, *The Social History of Lighting* (Routledge and Kegan Paul, 1958), p. 117. See also Percy Dunsheath, *A History of Electrical Engineering* (Faber and Faber, 1962), p. 141.

69. The arc lamps were designed by C. F. Brush, and the filament lamps by (Sir) Joseph Swan. See Smith, in *A History of Technology*, ed. Singer et al., vol. 5, p. 519. Also see C. Mackechnie Jarvis, "The Distribution and Utilization of Electricity," ibid., vol. 5, pp. 213–215.

70. Smith, in Singer et al., vol. 5, p. 500.

71. A. R. Griffin, *The Collier* (Shire Album 82, 1982), p. 19; O'Dea, pp. 123–124.

72. O'Dea, p. 123; Griffin, pp. 19, 22.

73. Information from Gordon L. Smith, cave historian and owner of Marengo Caverns, Indiana (private conversation, March 17, 1988).

74. *Illustrated London News*, September 22, 1849, quoted in Pointon, p. 89.

75. William Stump Forwood, *An Historical and Descriptive Narrative of the Mammoth Cave of Kentucky* (Philadelphia: J. B. Lippincott, 1870), pp. 58, 130, 156–159. An 1891 guidebook describes somewhat more precisely the type of lamp in use then: "a simple affair for burning lard-oil, [which] swings from four wires twisted into a handle, with a tin shield to protect the hand" (Horace C. Hovey, *Guide Book to the Mammoth Cave of Kentucky* [Cincinnati: Robert Clarke & Co., 1891], p. 16). The cavern tour also featured a "bottomless pit," which Knox notes was actually 175 feet deep (p. 470).

76. James W. Carey and John J. Quirk, *American Scholar* 39, no. 2 (spring 1970): 219–240; no. 3 (summer 1970): 395–424.

77. Ibid., no. 3, p. 395. Carey and Quirk pointedly use Leo Marx's term "the rhetoric of the technological sublime," rather than referring simply to technological sublimity.

78. Ibid., p. 396.

79. Thomas P. Hughes, "The Industrial Revolution That Never Came," *American Heritage of Invention and Technology* 3, no. 3 (winter 1988): 58–64. This article is adapted from a paper read before the First International Congress on the History of Electricity in France (1986) and published in *Un siècle d'électricité dans le monde* (Association pour l'Histoire de l'Électricité en France).

80. Hughes, p. 64.

81. I am borrowing the expression from Paul Goldberger's article "A Serene Place to Work, Not a Corporate Spaceship," *New York Times,* February 21, 1988.

82. Benjamin, *Baudelaire,* p. 171. The volume consists of three essays: "The Paris of the Second Empire in Baudelaire," "Some Motifs in Baudelaire," and "Paris—the Capital of the Nineteenth Century." "Motifs" is also found in Benjamin's *Illuminations,* ed. Hannah Arendt, tr. Harry Zohn (Harcourt, Brace & World, 1955). For more examples and analyses of surreal cities, see Inch, "Fantastic Cities," pp. 117–121.

83. Roger Caillois, *Le Mythe et l'homme* (Gallimard, 1972), pp. 154–159.

84. Benjamin, *Baudelaire,* p. 55.

85. Ibid., p. 166.

86. Ibid., p. 59.

87. Ibid., p. 37.

88. Ibid., p. 50.

89. Ibid., pp. 54, 158. Baudelaire's obsession with interiority has been analyzed more recently by T. J. Clark, who writes of the poet's "metaphysic of the city" as "something very like a religion"—the religion of artifice. "In the city," he notes, "Nature is absent and has to be invoked, conjured, recalled by the refinements of artifice." (*The Absolute Bourgeois: Artists and Politics in France 1848–1851* [Princeton University Press, 1982 (1973)], p. 175.)

90. Clark, p. 175.

91. Ibid.

92. J. W. Oliver, *The Life of William Beckford* (Oxford University Press, 1932), pp. 89–91; quoted in Gemmett, *William Beckford,* pp. 84–85.

93. Gemmett, p. 85.

94. Ibid.

95. These technological utopias amply demonstrate Arthur C. Clarke's Third Law: "Any sufficiently advanced technology is indistinguishable from magic" (*Report on Planet Three and Other Speculations* [Signet, 1973], p. 130). Clarke adds the important qualifier that technology is indistinguishable from magic only to those who do not understand the technology; he gives the hypothetical example of Edison confronted by the laser.

96. In other words, the rhapsodies of "overt culture" persist, but they are modified and to some extent contradicted by the language of "covert culture." For a discussion of these terms, see Leo Marx, "Literature, Technology, and Covert Culture (with Bernard Bowron and Arnold Rose)," *Pilot and Passenger,* pp. 127–138 (first published as "Literature and Covert Culture" in *American Quarterly,* 1957).

97. I am summarizing here the argument I made in "Corrupting the Public Imagination," *Christian Science Monitor,* March 20, 1981.

98. Benjamin, *Baudelaire,* p. 59.

99. Quoted in Jennings, *Pandaemonium,* pp. 187–188; from John W. Oliver's *Life of Beckford* (1932). The letter was originally written in French; the translation is mine.

100. Feaver, *Art of John Martin,* p. 62.

101. I am borrowing Raymond Williams's key term "structure of feeling" as described in his essay "Literature and Sociology: In Memory of Lucien Goldmann," *Problems,* pp. 22–27 (first published in *New Left Review* 67 [May-June 1971]; based on a lecture given in Cambridge in April 1971).

One obvious way to withdraw from ravaged nature into a (bio)technology-created paradise is through drugs. Drug use is a topic that deserves serious attention from cultural as well as from social historians. I cannot give it that attention here, but I do want to note its importance and its relevance to the issues raised here. According to Herbert Marcuse, "even the psychedelic withdrawal may be seen as creating 'artificial paradises within the society from which it withdrew'" (from Marcuse's *Essay on Liberation* [1969], quoted by Marx in "Susan Sontag's 'New Left' Pastoral," *Pilot and Passenger,* p. 296).

102. "William Beckford," *Encyclopaedia Britannica,* ninth edition (1878).

103. Quoted in Benjamin, *Baudelaire,* p. 94.

104. The closest approximation to such a study, which appropriately enough takes its title from Villiers's drama, is still Edmund Wilson's *Axel's Castle: A Study in the Imaginative Literature of 1870–1930* (Scribner, 1947).

105. Two English translations have appeared in recent years: *Eve of the Future Eden,* tr. M. Rose (Coronado, 1981), and *Tomorrow's Eve,* tr. R. Adams (University of Illinois Press, 1982). All quotations here are from the latter.

106. In the words of his biographer A. W. Raitt, "It is hard to know whether his distant and illustrious forebears or the more recent lineage of impoverished and eccentric Breton nobles had the greater influence on him." (*The Life of Villiers de l'Isle-Adam* [Clarendon, 1981], p. 3.) Another useful but much shorter biography is that by William T. Conroy, Jr., *Villiers de l'Isle-Adam* (Twayne, 1978).

107. *Tomorrow's Eve,* p. 7.

108. Ibid., p. 3.

109. Eugene S. Ferguson, review of *Edison's Electric Light: Biography of an Invention,* by Robert Friedel and Paul Israel with Bernard S. Finn, *Technology and Culture* 29, no. 1 (January 1988), p. 153.

110. Ibid.

111. *Tomorrow's Eve,* p. 18.

112. Ibid., pp. 10–11.

113. Ibid., p. 31.

114. Ibid., p. 44.

115. Ibid., p. 55.

116. Ibid., p. 71.

117. Ibid., p. 88.

118. Ibid., p. 91.

119. Ibid., pp. 92–93.

120. Ibid., pp. 97–98.

121. Ibid., p. 157.

122. Ibid., p. 165.

123. Ibid., p. 219.

Chapter 5

1. Camille Mauclair, "Villiers de l'Isle-Adam," *La Revue* 67, fourth series, no. 8 (April 15, 1907), p. 504.

2. Irving Howe, "Céline: The Sod Beneath the Skin," in *Decline of the New* (Harcourt, Brace & World, 1970), p. 54 (the essay first appeared in *A World More Attractive,* published by Horizon in 1963). See also Howe's description of the underground man in "The City in Literature," in *The Critical Point on Literature and Culture* (Horizon, 1973), p. 50 (first published in *Commentary* in 1971).

3. Tuan, *Man and Nature*, p. 34.

4. From Mill's essay in Coleridge, quoted in Gunn, p. 11. For a discussion of Mill's use of the degeneracy theory, see Stuart C. Gilman, "Political Theory and Degeneration: From Left to Right, From Up to Down," in Chamberlin and Gilman.

5. Eric T. Carlson, "Medicine and Degeneration: Theory and Praxis," in Chamberlin and Gilman, p. 122.

6. Ibid., p. 140.

7. Quoted in Mark R. Hillegas, "Cosmic Pessimism in H. G. Wells's Scientific Romances," *Papers of the Michigan Academy of Sciences, Arts, and Letters* 46 (1961): 658. For a review of various books on degeneracy, see Marquis de Castellane, "Les maladies du siècle," *Nouvelle revue* (4e livraison) 90 (October 15, 1894), pp. 807–818.

8. Richard Hauer Costa, *H. G. Wells* (Boston: Twayne, 1985), pp. 5–8.

9. These writings are indexed in *H. G. Wells: Early Writings in Science and Science Fiction*, ed. R. M. Philmus and D. Y. Hughes (University of California Press, 1975).

10. H. G. Wells, *The Time Machine* in *Seven Science Fiction Novels* (New York: Dover Publications, 1895), p. 26.

11. Ibid., p. 27.

12. Ibid., p. 28.

13. Ibid., p. 39.

14. Irving Howe, "The Fiction of Antiutopia," in *Decline*, p. 67 (first published in *A World More Attractive*, 1963).

15. Sussman, *Victorians and the Machine*, p. 174.

16. Costa describes Wells's relationship with the Fabians as "destructive to both the creative artist and the man" (p. 50).

17. Stephen Jay Gould, *The Mismeasurement of Man* (Norton, 1981), especially p. 80.

18. Raymond Williams, "Social Darwinism," in *Problems*, p. 86. This article is based on a lecture given at the Institute of Contemporary Arts in London in 1972, which was published in *The Limits of Human Nature*, ed. J. Benthall (Allen Lane, 1973).

19. From *Grundrisse, Foundations of the Critique of Political Economy*, cited by Schivelbusch, p. 164. See also the analysis by Sidney Hook in *From Hegel to Marx* (University of Michigan Press, 1962), p. 277: "Human history may be viewed as a process in which new needs are created as a result of material changes instituted to fulfill the old. According to Marx . . . the changes in the character and quality of human needs, including the means of gratifying them, is the keystone not merely to historical change but to the changes of human nature."

20. See Irving Howe's comments on the "indestructible core" of plastic human nature in "The Fiction of Antiutopia," p. 70.

---

21. Bulwer-Lytton, pp. 118–119. Born Edward George Earle Lytton Bulwer, the author of *The Coming Race* assumed his mother's surname upon her death in December 1843 and became Edward Bulwer-Lytton. In July 1866 he assumed the title of Baron Lytton of Knebworth and was thereafter called Lord Lytton. For the sake of simplicity, I shall refer to him consistently as Bulwer-Lytton and to his wife as Lady Lytton. For biographical details see the article (listed under *Lytton*) in *Dictionary of National Biography,* ed. Sidney Lee (Macmillan; Smith and Elder, 1893), vol. 34, pp. 380–387.

22. Letter dated June 1871; quoted in Campbell, p. 126.

23. Bulwer-Lytton, p. 237.

24. Ibid., p. 238.

25. Williams, "Utopia and Science Fiction," in *Problems,* p. 201.

26. Susan Sontag, "The Imagination of Disaster," in *Against Interpretation and Other Essays* (Farrar, Straus & Giroux, 1961), p. 221.

27. Ibid.

28. Bulwer-Lytton died in January 1873.

29. Campbell, p. 65.

30. In *The Coming Race* the choice of an American narrator also permits Bulwer-Lytton to satirize the ill effects of leveling democracy in the United States, where the citizens, "accustomed from infancy to the daily use of revolvers, should apply to a cowering universe the doctrine of the Patriot Monroe" (pp. 43–44). This passage is quoted by Geoffrey Wagner in "A Forgotten Satire: Bulwer-Lytton's *The Coming Race,*" *Nineteenth Century Fiction* 19 (March 1965): 379–385. Wagner seems to overemphasize the United States as a special object of satire in the book.

31. Bulwer-Lytton, pp. 227–228.

32. Ibid., pp. 70–71.

33. Ibid., pp. 114–115.

34. Ibid., pp. 247–248.

35. Campbell, p. 18.

36. E. M. Forster, "The Machine Stops," in *Collected Stories,* p. 125.

37. Ibid., p. 148.

38. The most elaborate and influential example of this treatment is found in the "Book of the Machines" in Samuel Butler's *Erewhon,* published in 1872 (shortly after *The Coming Race* appeared). The satires were similar enough that in his preface to the second edition Butler took pains to show that he had held his ideas for a long time, and that he had not read Bulwer-Lytton's book until his own was completed. See Sussman, p. 145; Wagner, pp. 379–380.

39. Forster, p. 115.

40. As Virginia Woolf said of Forster's writing in general, he keeps telling us that the disease is convention and nature is the remedy. See discussion by Wilfred Stone in *The Cave and the Mountain: A Study of E. M. Forster* (Stanford University Press, 1966), p. 155.

41. Forster, p. 157.

42. Ibid., pp. 140–141.

43. Ibid., p. 133.

44. This childbirth imagery is analyzed by Stone (pp. 152–155). Stone also notes the significance of the safe, enclosed dell in Forster's work, notably the Cambridge dell in *The Longest Journey* (1907).

    I am avoiding the complexities of interpreting Forster's best-known subterranean image, the Marabar Caves in *A Passage to India* (1924). For a summary of the various interpretations of these caves, see Frederick P. W. McDowell, *E. M. Forster* (Twayne, 1969), pp. 105–111.

45. Forster, p. 143. As Stone shows (p. 155), Forster's language quite clearly indicates that Kuno is castrated by the Mending Apparatus in this episode. For another, less favorable reading of the tale, see Frederick P. W. McDowell, "Forster's 'Natural Supernaturalism': The Tales," *Modern Fiction Studies* 7 (1961), pp. 271–283.

46. Forster, p. 144.

47. Ibid., p. 152.

48. Ibid., p. 156.

49. Ibid., pp. 156–158.

50. For more biographical details, see Williams, *Dream World: Mass Consumption in Late-Nineteenth-Century France* (University of California Press, 1982), pp. 342–346; Terry N. Clark, "Gabriel Tarde," in *International Encyclopedia of the Social Sciences,* ed. David L. Sills (1968), vol. 15, pp. 509–514; and Jean Milet, *Gabriel Tarde et la philosophie de l'histoire* (Librairie Philosophique J. Vrin, 1970).

51. Tarde, pp. 111–113.

52. Ibid., p. 113.

53. Ibid., p. 141.

54. Ibid., p. 146.

55. Clark, p. 175.

56. Tarde, p. 122.

57. Ibid., pp. 86–88.

58. Ibid., p. 105.

59. Ibid., pp. 119–120.

60. Ibid., p. 106.

61. Ibid., pp. 150–151.

62. Ibid., p. 151.

63. Ibid., pp. 152–153.

64. Ibid., pp. 156–160.

65. Ibid., pp. 123–124.

66. Orwell, *Wigan Pier,* pp. 169–170. See also the comments on this passage in Hillegas, *Future as Nightmare,* p. 127.

67. Orwell, p. 172.

68. Forster, p. 119.

69. Bulwer-Lytton, pp. 230–231.

70. Howe, in "Fiction of Antiutopia," notes that the most intense antiutopian fiction comes from "men of the left" (Yevgeny Zamyatin, George Orwell, Aldous Huxley) who discover themselves "struck with horror" when they contemplate a future they have been "trained to desire" and turn "upon their own presuppositions" (p. 67).

71. Orwell, p. 184.

72. Ibid., p. 188.

73. I regret that I do not have the space to analyze another such subterranean social environment: Herbert Read's *The Green Child: A Romance* (London: William Heinemann, 1935). It is well worth reading not only for its intrinsic merit, but also for the way it illuminates the themes discussed here. Read was a philosophical anarchist, and any interpretation of *The Green Child* would have to relate it to his political convictions, as well as to explore its biographical and archetypal resonances.

For more information on Read and *The Green Child,* see George Woodcock, *Herbert Read: The Stream and the Source* (Faber and Faber, 1972), especially pp. 33–39 and 66–77; Robert Melville, "The First Sixty-Six Pages of *The Green Child,*" in *Herbert Read: An Introduction to His Work by Various Hands,* ed. H. Treece (Faber and Faber, 1944), pp. 81–90; and *Herbert Read: A Memorial Symposium,* ed. R. Skelton (Methuen, 1970).

74. Howe, "The City in Literature," p. 45.

75. Mumford, *Technics,* p. 280.

76. Bulwer-Lytton, p. 224.

77. Ibid., pp. 57, 161.

78. Ibid., p. 186.

79. Ibid., p. 59.

80. Tarde, pp. 131–132.

81. Ibid., p. 151.

82. Bulwer-Lytton, p. 230.

83. Ibid., p. 220. Compare this passage with Leo Marx's analysis of Hester Prynne's plea in "The Puzzle of Anti-Urbanism in Classic American Literature,"

in *Pilot and Passenger,* pp. 218–220 (first published in *Literature and the Urban Experience, Essays on the City and Literature,* ed. Michael C. Jaye and Ann Chalmers Watt [Rutgers University Press, 1981]).

84. Marx, "Puzzle," p. 210.

85. Forster, p. 138.

86. Ibid., p. 143.

87. Ibid., p. 140.

88. Ibid., p. 141.

89. Ibid., p. 157.

90. Bodenheimer, *Politics of Story,* p. 115. Bodenheimer gives an impressive political interpretation of pastoralism in part 2, "The Pastoral Argument."

91. Daniel Callahan, *The Tyranny of Survival and Other Pathologies of Civilized Life* (Macmillan, 1973), p. 39.

92. Forster, pp. 129–130.

93. Ibid., p. 129.

94. Gordon A. Craig, *Europe since 1815,* second edition (Holt, Rinehart and Winston, 1966), p. 395.

95. Spengler, *Decline,* vol. 2, p. 90.

96. Williams, "Base and Superstructure in Marxist Cultural Theory," in *Problems,* p. 41.

97. Quoted by Eric S. Rabkin, in "Introduction: Why Destroy the World?" in *The End of the World,* ed. E. S. Rabkin, M. H. Greenberg, and J. D. Olander (Southern Illinois University Press, 1983), p. xi.

98. See Arendt, *Human Condition,* passim.

99. Bulwer-Lytton, pp. 114, 171.

100. Forster, p. 121.

101. Ibid., p. 155.

Chapter 6

1. Asa Briggs, "The Language of 'Class' in Early Nineteenth Century England," in *Essays in Labour History,* ed. A. Briggs and J. Saville, 1960; cited in P. J. Keating, *The Working Classes in Victorian Fiction* (Routledge & Kegan Paul, 1971), p. 10.

2. Marx, *Machine,* p. 187.

3. The phrase is quoted by Klingender in *Art in the Industrial Revolution,* p. 171.

4. In 1836 Peter Gaskell (later considered by Marx and Engels as the leading British authority in industrial medicine) summarized the health of industrial

workers in this way: "On the whole, it may be said that the class of manufacturers engaged in mill labour, exhibit but few well-defined diseases; but that nearly the entire number are victims to a train of irregular morbid actions, chiefly indicated by disturbances in the functions of the digestive apparatus, with their consequent effects upon the nervous system; producing melancholy, extreme mental irritation, and great exhaustion." (Gaskell, *Artisans and Machinery: The Moral and Physical Condition of the Manufacturing Population Considered with Reference to Mechanical Substitutes for Human Labour* [1836], quoted in Schivelbusch, *Railway Journey*, p. 119)

5. See, for example, the transcripts collected by Pike (ed.), *"Hard Times": Human Documents of the Industrial Revolution.* See also the documents under the headings "Industrial Man" and "Daemons at Work" in Jennings, *Pandaemonium.*

6. See also "The Nether World: Class and Technology," in Lesser, *Life Below the Ground*, pp. 77–101.

7. Bodenheimer, *Politics of Story*, p. 3.

8. Quoted in Jennings, pp. 32—36. More than 250 years later, some of the poor still live literally underground. New York City transit officials estimate that at least several hundred homeless people live in the 700 miles of subway tunnels below the city streets ("Life in the Underworld: When a Tunnel is Home," *New York Times,* January 1, 1989).

9. Keating (*Working Classes in Victorian Fiction,* pp. 130, 138) uses this metaphor.

10. Charles Dickens, *The Pickwick Papers* (New American Library, 1964), pp. 437–447.

11. Hugo, pp. 718–727.

12. Ibid., p. 997.

13. Ibid., p. 1123.

14. Heralded by Harriet Martineau's *A Manchester Strike* (1832), the best-known examples are Mrs. Trollope's *Michael Armstrong* (1839–40), Disraeli's *Coningsby* (1844) and *Sybil* (1845), Mrs. Gaskell's *Mary Barton* (1848) and *North and South* (1855), Charles Kingsley's *Alton Locke* (1850), Dickens's *Hard Times* (1854), and George Eliot's *Felix Holt* (1866). See Keating, pp. 5–9, and Bodenheimer, pp. 4–6.

15. Quoted by Keating, p. 20.

16. Keating, pp. 6–8, 23, 224.

17. Raymond Williams, *Culture and Society 1780–1950* (Harper & Row, 1958), p. 109.

18. Keating, p. 106.

19. Ibid., p. 124.

20. Quoted in Keating, p. 67; see also p. 136.

21. Quoted in Keating, p. 85.

22. Quoted in Keating, p. 92. See also Raymond Williams's perceptive discussion of Gissing's identification with the poor in *Culture and Society*, pp. 172–179.

23. Keating, pp. 117–119.

24. Such an accidental and unfortunate descent into the social underworld is the premise of Tom Wolfe's novel *The Bonfire of the Vanities* (1987). Wolfe vigorously defends the realistic-naturalistic tradition, which he feels has been discarded far too carelessly by twentieth-century writers. Wolfe further insists on the importance of doing detailed research in preparation for realistic writing. He cites the example of Zola descending into working mines in preparation for writing *Germinal*. There Zola discovered the plight of mine ponies who spent their entire lives underground—a detail he used in his novel as a metaphor for the plight of the human miners. "I insist," Wolfe said, "that the only way a writer can come up with material like that is to do what Zola did, which is to plunge into the life of the society around him. . . ." (interview by Gisela M. Freisinger, *Spin* 4, no. 7 [October 1988], pp. 62–63, 72–73)

25. Michael Wilding, *Political Fictions* (Routledge and Kegan Paul, 1980), pp. 5–6.

26. Irving Howe, *Politics and the Novel* (London: Stevens, 1961 [1957]), p. 19; quoted in Wilding, p. 3.

27. Wilding, pp. 5–6.

28. Ibid., p. 8.

29. Verne, p. 279.

30. Ibid., p. 280.

31. Ibid., pp. 281–282.

32. Ibid., p. 299.

33. Ibid., pp. 318–319.

34. Jean Jules-Verne, *Jules Verne*, pp. 135–136.

35. Verne, *Black Indies*, p. 341.

36. Ibid., p. 295.

37. In his influential Marxist interpretation of Verne, Pierre Macherey has described this theme of colonization—constructing a new America in *The Mysterious Island,* or a new Aberfoyle in *Black Indies*—as not precisely a utopian vision, because for Verne it is not a matter of inventing new conditions but of reproducing what is possible given certain conditions. (Macherey, *Pour une théorie de la production littéraire* [Paris: François Maspero, 1966], pp. 241–242)

38. Chesneaux, pp. 61–76.

39. Verne, *Black Indies,* p. 394.

40. Ibid., p. 344.

41. Ibid., p. 383.

42. Ibid., p. 389.

43. Ibid., pp. 391–393.

44. Ibid., p. 389.

45. J. Jules-Verne, p. 215.

46. Chesneaux, p. 186. For more comparisons between Villiers and Verne, see Marcel Moré, *Le très curieux Jules Verne* (Gallimard, 1960). Moré points out the striking resemblances between Villiers's *L'Ève future* and Verne's *Le Château des Carpathes* [The Castle of the Carpathians]. Verne's book was published in 1892, but he had been working on *Château* as early as 1889. Moré speculates that Verne was influenced by *L'Ève* as he made final corrections on his own manuscript. Moré, however, incorrectly gives 1891 as the date that *L'Ève* was published. It in fact appeared in 1886, so Verne might well have read it even before beginning *Château*. See J. Jules-Verne, pp. 173–176, for a plot summary and for comparisons between the two works.

47. Verne, *Twenty Thousand Leagues*, pp. 74–75.

48. Ibid., p. 94.

49. Ibid., p. 376.

50. Ibid., p. 393.

51. Ibid., p. 331.

52. Chesneaux, pp. 167–169. The remark about Lang is made with reference to Blackland, which appears in *L'Étonnante Aventure de la Mission Barsac*, Verne's last book (it was published posthumously in 1914, after editing by Verne's son Michel). Other examples of "cities of perdition" cited by Chesneaux are Stahlstadt, the city of steel in *Les Cinq cents millions de la Begum* (1879), and Milliard City in *L'île à hélice* (1895).

53. Costa, *H. G. Wells*, p. 105. Costa credits Hillegas's book *Future as Nightmare* for winning general recognition for the dystopian influence of "Wellsian" literature (p. 139).

54. Anthony West, Wells's son, wrote in 1967 that his father was "by nature a pessimist, and he was doing violence to his intuitions and his rational perceptions alike when he asserted in his middle period that mankind could make a better world for itself by an effort of will." (*H. G. Wells: A Collection of Critical Essays*, ed. B. Bergonzi [Prentice-Hall, 1976], p. 10)

55. In 1899 Wells also published *A Story of the Days to Come* (it appeared in his *Tales of Space and Time*), which is set in the same future world but at a slightly later time (*Sleeper* was supposed to have taken place in the spring of A.D. 2100; *Story* begins in late 2100 and extends to about 2103 or 2104). *Story* has the same vertical arrangement of the upper classes living "in a vast series of sumptuous hotels in the upper storeys and halls of the city fabric," while "the industrial population dwelt beneath in the tremendous ground-floor and basement, so to speak, of the place" (Wells, *A Story of the Days to Come*, in *Three Prophetic Novels*, ed. E. F. Bleiler [Dover, 1960], p. 209). The plot concerns a young couple forced

by circumstances to live "Underneath" (the title of the fourth chapter) until they are rescued from their miserable existence by a fortunate legacy. The most memorable part of *Story* is the mock-pastoral narrative of the lovers' initial attempt to flee the great city and live in the country (chapter 2, "The Vacant Country")—a sad and moving elegy for the possibility of a pastoral retreat.

56. Wells, *Time Machine*, pp. 34–35.

57. Ibid., p. 41.

58. Ibid., pp. 41–42.

59. Ibid., p. 51.

60. Ibid., p. 65.

61. H. G. Wells, *When the Sleeper Wakes*, in *Three Prophetic Novels* (Dover, 1960), pp. 28–29.

62. Paul M. Jensen, *The Cinema of Fritz Lang* (London: Zwemmer, 1969), pp. 7, 62. Wells's comment was made in "Mr. Wells Reviews a Current Film," *New York Times*, April 17, 1927. On *Sleeper* as a source for *1984*, see Wilding, pp. 217–221.

63. Wells, *Sleeper*, p. 29.

64. Ibid., p. 55.

65. Ibid., p. 137.

66. Ibid., pp. 155–156.

67. Ibid., p. 169.

68. Ibid., pp. 173–174.

69. Ibid., p. 175.

70. Ibid., p. 187.

71. Preface to an edition published by W. Collins Sons & Co., London (no date), p. 6.

72. Wells, *Sleeper*, p. 38.

73. Ibid., p. 137.

74. In his preface to the revised edition of *Sleeper*, Wells states that he eliminated "certain dishonest and regrettable suggestions that the People beat Ostrog." He continues: "My Graham dies, as all his kind must die, with no certainty of either victory or defeat." (quoted in Sussman, *Victorians and the Machine*, p. 188)

75. In his critique of *Sleeper*, George Orwell pointed to a basic inconsistency in Wells's view of the future: ". . . in the immensely mechanized world that Wells is imagining, why should the workers have to work harder than at present? Obviously the tendency of the machine is to eliminate work, not to increase it. In the machine-world the workers might be enslaved, ill-treated, and even underfed, but they certainly would not be condemned to ceaseless manual toil; because in that case what would be the function of the machine? You can have

machines doing all the work or human beings doing all the work, but you can't have both." (Orwell, *Wigan Pier,* pp. 177–178)

76. Wells, preface to *Sleeper,* p. 6. In this preface Wells says that in writing the book he overestimated the intelligence and power of capitalist bosses. Having met more businessmen since then, Wells says, he has concluded they are "for the most part, rather foolish plungers, fortunate and energetic rather than capable, vulgar rather than wicked, and quite incapable of world-wide constructive plans or generous combined action."

77. Wells, *Sleeper,* p. 236.

78. H. G. Wells, *The First Men in the Moon,* in *Seven Science Fiction Novels of H. G. Wells* (Dover, 1901), p. 512.

79. Ibid., p. 555.

80. Ibid., p. 604.

81. Ibid., p. 601.

82. Ibid., p. 606.

83. Ibid., pp. 604–605.

84. Ibid., p. 609.

85. This combination of hypercerebrality and bodily withering is reminiscent of Vashti. It is evident, from numerous similarities in language and theme, that Forster was influenced by *The First Men in the Moon* as well as *The Time Machine* in composing "The Machine Stops."

86. The contradiction between Marxist predictions and the lunar society of *First Men in the Moon* was first noted by George Connes in *Études sur la pensée de Wells* (1926)—see Costa, p. 27.

87. Marx, "Susan Sontag's 'New Left' Pastoral," in *Pilot and Passenger,* p. 301.

88. The economist and social critic Albert O. Hirschman, in *Shifting Involvements: Private Interest and Public Action* (Princeton University Press, 1982), suggests that American history reveals a pattern of alternating (but not simultaneous) attention to the private and the public spheres.

89. Marx, "Sontag," p. 293.

90. Clark, *Absolute Bourgeois,* p. 176.

91. Marx, "Sontag," p. 292.

92. Fred Hirsch, *Social Limits to Growth* (Harvard University Press, 1976).

93. Howe, "City in Literature," p. 57.

Chapter 7

1. Howe, "The City in Literature," *Critical Point,* p. 54.

2. In *Terminal Visions* (Indiana University Press, 1982), W. Warren Wagar analyzes the apocalyptic prophecies of more than 300 novels, stories, plays, and poems written since the late 1800s in what he calls the "Anglo-Franco-American heartland." I. F. Clarke has gathered predictions of military doom from the same era in *Voices Prophesying War, 1763–1984* (Oxford University Press, 1966). See also Clarke's *The Pattern of Expectation: 1644–2001* (Basic Books, 1979).

3. In his study of French entertainments in the Belle Epoque, Charles Rearick adjusts a common interpretation of that age as "a last fling before a long antic-ipated cataclysm" by noting that "in many popular entertainments people were not so much escaping dread as they were finding danger to be a part of their enjoyment." In the various shows and amusements he describes, fear and horror could be experienced "without having to bear the real costs and pain." (Charles Rearick, *Pleasures of the Belle Epoque: Entertainment and Festivity in Turn-of-the-Century France* [Yale University Press, 1985], pp. 202, 207)

John Kasson finds the same taste for sublimity in American entertainments of that epoch. At Coney Island, visitors could see simulations of the fall of Pompeii, the eruption of Mount Pelee in 1902, and the Johnstown flood of 1889. "Such displays," Kasson writes, "reflected a fascination with disaster . . . a horrible delight in the apprehension that devastating tragedy had . . . intruded suddenly in daily affairs, even in modern technological America." (John Kasson, *Amusing the Million: Coney Island at the Turn of the Century* [Hill and Wang, 1978], pp. 71–72.) In his earlier book *Civilizing the Machine,* Kasson had stressed the popular craving for the delight, awe, and fear aroused by steam engines, locomotives, and other powerful machines. These devices aroused sublime feelings as an unintentional side effect; the Coney Island machines had this as their primary purpose.

Some of the earliest mechanical amusement rides were consciously designed to convey the sensation of descending underground. The Switchback Railroad, initiated in 1884 and a forerunner of the roller coaster, was modeled after the cars used by coal miners to descend into the pit. Later a mechanized version of this ride was devised that extended the mining analogy by having the cars run through a tunnel (*Amusing,* p. 74).

4. For discussion of "the democratization of luxury," see Williams, *Dream Worlds,* especially pp. 213–233.

5. This phrase is used by Carl Smith in "Urban Disorder and the Shape of Belief: The San Francisco Earthquake and Fire," *Yale Review,* autumn 1984, p. 92.

6. See Wagar, *Terminal Visions,* pp. 33–53.

7. Frank Kermode argues that contemporary fiction and drama (culminating in absurdist fiction) represent a sophisticated, personalized version of ancient apoc-alyptic prophecy (*The Sense of an Ending: Studies in the Theory of Fiction* [Oxford University Press, 1967]). Science fiction has often been interpreted as a distinc-tively modern version of apocalyptic literature; see Robert Galbreath, "Ambig-

uous Apocalypse: Transcendental Versions of the End," in *The End of the World,* ed. E. Rabkin, M. Greenberg, and J. Olander (Southern Illinois University Press, 1983), p. 56. Galbreath stresses the importance of David Ketter's book *New Worlds for Old: The Apocalyptic Imagination, Science Fiction, and American Literature* (Doubleday Anchor, 1974). Ernest Lee Tuveson theorizes that millennialism combined with Newtonian physics to emerge as the modern belief in progress (*Millennium and Utopia: A Study in the Background of the Idea of Progress* [University of California Press, 1949]). Even more daringly, Norman Cohn has argued that the mode of secularization was not so much literary or intellectual as political. His thesis is that modern totalitarianism (both the Nazi and the Soviet variety) is a direct descendent of the militant, revolutionary chiliasm of the late Middle Ages and the early Reformation: "[That peculiar faith] continued a dim, subterranean existence down the centuries, flaring up briefly in the margins of the English Civil War and the French Revolution, until in the course of the nineteenth century it began to take on a new, explosive vigour, now in France, now in Germany, now in Russia." (Cohn, *The Pursuit of the Millennium: Revolutionary Messianism in Medieval and Reformation Europe and Its Bearing on Modern Totalitarian Movements,* second edition [Harper Torchbooks, 1961], p. 309)

8. For a discussion of Freud's theories in relation to visions of catastrophe, see Daniel Callahan, *The Tyranny of Survival and Other Pathologies of Civilized Life* (Macmillan, 1973), pp. 23–53.

9. George Steiner, *In Bluebeard's Castle: Some Notes towards the Redefinition of Culture* (Yale University Press, 1971), pp. 11, 20.

10. Ibid., p. 24.

11. Ibid., p. 19.

12. Such assumptions also pervade Spencer Weart's more recent study of nuclear imagery. According to Weart, who writes from a Jungian perspective, the human psyche contains archetypes that are expressed in an ancient cluster of images centered on the concept of transmutation. Those images were reified in the twentieth-century discovery of atomic energy: "Nuclear energy, with its wealth of ambiguous associations, served well as a receptacle for projection of these hidden thoughts. Impossible hopes could seem almost plausible when attached to possession of bombs and reactors. . . . People have not only projected their feelings onto bombs and reactors but have built these devices purposely to be what they are, turning visionary fires into real ones. Our secret thoughts have come into the open at last, taking form in metal so that we can deny them no longer." (Weart, *Nuclear Fear: A History of Images* [Harvard University Press, 1988], pp. 424–425)

13. Sontag, "The Imagination of Disaster," in *Against interpretation,* p. 224.

14. Wagar notes (but does not elaborate upon) an interesting national difference here: "In the two hundred and fifty works inventoried by this writer, the ratio of natural to man-made disasters in American fiction is approximately one to two, and in British and European fiction, one to one." (note 41 to Wagar, "The Rebellion of Nature," in Rabkin et al., p. 185)

15. Yi-Fu Tuan, *Space and Place: The Perspective of Experience* (University of Minnesota Press, 1977), p. 107.

16. Tuan, Topophilia, p. 118.

17. Ibid. Giedion deals with the transfer of prehistoric experiences and traditions to the civilizations of Egypt and Sumer in *The Eternal Present—The Beginnings of Architecture* (Pantheon, 1964).

18. Gould, *Time's Arrow,* pp. 121, 129.

19. Charles Lyell, *Principles of Geology,* volume 1, p. 9 (1830), quoted in ibid., p. 122.

20. Wagar, "Rebellion," in Rabkin et al., p. 144. For the scientific developments, see D. L. S. Cardwell, *From Watt to Clausius: The Rise of Thermodynamics in the Early Industrial Age* (London: Heinemann, 1971). For the diffusion of these ideas into other intellectual fields, see Stephen Brush, "Thermodynamics and History," *Graduate Journal* 7, no. 2 (1967), pp. 477–565. Also see G. J. Whitrow's article on "Entropy" in *Encyclopedia of Philosophy* (volume 2, pp. 526–529) and Max Jammer's in *DHI* (volume 2, pp. 112–121, especially the last section on "Extrascientific Consequences").

21. Bulwer-Lytton, p. 52.

22. Ibid., pp. 52–53.

23. The same device of projecting memory on a television screen is used much more recently in René Barjavel's *La Nuit des temps* (1968), translated into English as *The Ice People.* Thanks to this device, the scientists who discover a lost city buried deep in the polar ice cap are able to watch its last days.

24. Chesneaux, pp. 172–173, quoting *L'Île mystérieuse,* pp. 194–195.

25. Chesneaux, quoting *Vingt mille lieues,* p. 271.

26. Verne, *Off on a Comet,* tr. E. Roth (Dover, 1960), p. 231.

27. Tarde, pp. 50–63 passim.

28. Flammarion was a professional astronomer and science writer; his *Astronomie populaire* (1880) "soon became one of the best-known books of its kind in the Western world." The first volume of his *La Planète Mars* appeared in 1892, and the descriptions of Mars in the first chapter of Wells's *The War of the Worlds* (1898) are strikingly similar to those in Flammarion's book. *La fin du monde* [Omega: The Last Days of the World] (New York: Cosmopolitan, 1894; New York: Arno, 1975) tells of the solar system 10 million years in the future, when cold and drought force humankind back into two cities of iron and glass situated along the Equator in the dry beds of what had been the Pacific and Indian oceans. Eventually all life on earth ends, but in the closing pages of the book Flammarion presents some spiritualist beliefs that suggest a new civilization will flourish on Jupiter. See Rabkin et al., pp. 62, 86–87, 151–152; Wagar, *Visions,* pp. 94, 177.

Hodgson's book (London: Nash, 1912; Westport, Conn.: Hyperion, 1976) is set several million years from now, when the sun no longer shines and the earth's surface no longer supports life. One hundred miles below the surface, in a huge crack caused by some earlier disaster, the last remnant of humankind lives in an eight-mile-high pyramid powered by a mysterious "Earth-Current" force. The Night Land around the pyramid is warmed by volcanic fires; it is inhabited by monstrously degenerate descendants of *Homo sapiens*. When the Earth-Current fails, as it must, the Pyramid will fall (Wagar, p. 94).

29. Wagar, "Rebellion," in Rabkin et al., pp. 149–150.

30. William Delisle Hay, *The Doom of the Great City; being the narrative of a survivor, written AD 1942* (London: Newman and Co., 1882), pp. 32, 36. I. F. Clarke comments on Hay in *Pattern*, p. 158.

31. Whether or not London's air quality was worsening during the nineteenth century, there is no question that it had severely deteriorated in the middle of the twentieth century. A disastrous smog in December 1952 was responsible for about 4,000 deaths in London. Steps to improve air quality were instituted, but in 1962 another December smog killed 340 people. See T. J. Chandler, *The Climate of London* (London: Hutchinson, 1965), p. 247. The main conclusions of Chandler's book are that "Londoners live in a profoundly man-modified climate" and that most of these modifications have been detrimental.

32. Verne, *Black Indies*, p. 291.

33. Ibid., pp. 291–292.

34. Wells, *Time Machine*, p. 71.

35. Wells, *The War of the Worlds*, in *Seven Novels*, pp. 434–435.

36. Tarde, pp. 104–105.

37. Verne, *Twenty Thousand Leagues*, p. 397.

38. In his structuralist reading of *Black Indies*, Michel Serres, a historian of science and technology, interprets the book as an "alchemy with four elements": the earth and the underground, lakes and lochs, the air space of the caverns, and the fire of the coal. The final explosion (or near-explosion) at the end of the book would result from this "elementary tetrology." Chesneaux criticizes Serres's reading for being overly static and psychological; in contrast, Chesneaux says, "our political analysis of the *Extraordinary voyages* has in effect led us to insist constantly on the presence, the omnipresence of the nineteenth century in that work—that work which belongs to its time" (p. 185). Serres's interpretation was published as "Un voyage au bout de la nuit" (*Critique*, no. 263 [April 1969]) and "Oedipe-messager" (no. 272 [January 1970]).

39. Michael Barkun, *Disaster and the Millennium* (Yale University Press, 1974), p. 51. The term "'true' society" is used by H. B. M. Murphy; "mazeway" is used by Anthony F. C. Wallace.

40. Ibid., p. 211.

41. Ibid., p. 184.

42. Ibid., p. 184.

43. Quoted by Kenneth Hewitt, in "The Idea of Calamity in a Technocratic Age," in *Interpretations of Calamity from the Viewpoint of Human Ecology*, ed. K. Hewitt (Allen & Unwin, 1983), p. iii; cf. p. 26. Hewitt comments that most

natural extremes are "more expected and knowable than many of the contemporary social developments that pervade everyday life" (p. 4).

44. Wells, *War,* p. 435.

45. Orwell, *Wigan Pier,* p. 170.

46. Barkun, p. 163.

47. Eric J. Hobsbawn, *Primitive Rebels: Studies in Archaic Forms of Social Movements in the Nineteenth and Twentieth Centuries* (Norton, 1959). See also the summary discussion in Barkun, pp. 182–185.

48. Sheldon Glashow, with Ben Bova, *Interactions: A Journey Through the Mind of a Particle Physicist and the Matter of This World* (Warner Books, 1988), pp. 305–306.

49. Sharon Traweek, *Beamtimes and Lifetimes: The World of High Energy Physicists* (Harvard University Press, 1988), pp. 36–37.

50. Thomas R. Kuesel, "Current Applications of Tunneling," in *Tunneling and Underground Transport: Future Developments in Technology, Economics, and Policy,* ed. F. P. Davidson (Elsevier, 1987), p. 23.

51. Ibid., p. 27.

52. Stu Campbell, *The Underground House Book* (Charlotte, Vermont: Garden Way, 1980), pp. 6–13.

53. Kenneth B. Labs, Undercurrent: The Architectural Use of Underground Space: Issues and Applications, master's thesis, 1975, School of Architecture, Washington University, St. Louis, p. II-10.

54. "Converted Mines Draw Businesses in Kansas City," *New York Times,* December 26, 1987.

55. Édouard Utudjian, with Daniel Bernet, presented by Michel Ragon, *Architecture et urbanisme souterrains* (Montreuil: Sociéte d'Impressions Publicitaires, 1966), p. 43.

56. Labs, p. II-18. Gunnar Birkerts comments on Doxiadis in *Subterranean Urban Systems* (University of Michigan, Industrial Development Division, Institute of Science and Technology, 1974).

57. Philip Deane, *Constantinos Doxiadis: Master Builder for Free Men* (Dobbs Ferry, N.Y.: Oceana, 1965), pp. 59–75.

58. Doxiadis, quoted in Labs, p. II-20.

59. Ibid.

60. Paolo Soleri, *Arcology: The City in the Image of Man* (MIT Press, 1969).

61. Paolo Soleri, *The Bridge between Matter and Spirit is Matter Becoming Spirit: The Arcology of Paolo Soleri* (Doubleday Anchor Books, 1973), p. 85.

62. Donald Wall, *Visionary Cities: The Arcology of Paolo Soleri* (Praeger, 1971), no page numbers.

63. Paolo Soleri, *The Sketchbooks of Paolo Soleri* (MIT Press, 1971), p. 8. Soleri stresses that earth houses are *semi*subterranean: "Human life is not mole life because human biopsychical characteristics are not those of the mole." Earth houses are not intended to be sealed off from nature but to be "amply open to sunlight, air, sounds, climate, and so forth" (p. 217).

64. Seiichi Kanise, "Japan's Underground Frontier," *Time*, February 6, 1989, p. 74.

65. Paul Fussell, *The Great War and Modern Memory* (Oxford University Press, 1975), pp. 36–74.

66. Mumford, *Technics*, p. 68.

67. Jeremy Errand, *Secret Passages and Hiding-Places* (David and Charles, 1974), pp. 181–192.

68. "Swiss Ready to Face Armageddon, in Comfort," *New York Times*, November 27, 1987.

69. Jim Robbins, "Visitors to a Small Planet: Earth, Meet Biosphere II," *New York Times*, October 18, 1987.

70. William J. Broad, "Ultimate Survival: Desert Dreamers Build a Man-Made World," *New York Times*, May 27, 1986.

71. Fussell, p. 74.

72. Kafka, "The Burrow," tr. W. and E. Muir, in *The Complete Short Stories*, ed. N. Glatzer (Schocken, 1976), p. 325ff.

73. Ernst Pawel, *The Nightmare of Reason: A Life of Franz Kafka* (Farrar, Straus and Giroux, 1984), p. 441.

74. As the geographer Tuan reminds us, humanity's increased technological power has two closely related effects: "to construct a man-made world of increasing complexity" and "to simplify nature's structured systems." "Both," he continues, "tend to make for instability." (*Man and Nature*, p. 42)

75. Callahan (*Tyranny of Survival*, pp. 127 and 267) stresses the importance of the ideas of Philip Rieff, who proposes a "release culture" as the means of increasing individual well-being at the expense of community life.

Afterword

Introduction: Five Bedrock Passages

There are five passages in *Notes on the Underground* where the reader, new or returning, touches the bedrock. The first is the opening sentence, which poses the question that motivated the book:

What are the consequences when human beings live in an environment that is predominantly built rather than given?

These consequences are psychological, social, political, and cultural as well as material. *Notes on the Underground* uses subterranean environments, real and imaginary, to consider the full range of these effects. In doing so, it traces the emergence in the late nineteenth century of what we now call environmental consciousness.

A second defining passage comes at the end of the book (pp. 212–213):

We have reached a point in human history where . . . we are preparing to descend below the surface of the earth forever. . . . We have always lived below the surface, beneath the atmospheric ocean, in a closed, sealed, finite environment, where everything is recycled and everything is limited. Until now, we have not felt like underground dwellers because the natural system of the globe has seemed so large in comparison with any systems we might construct. That is changing. What is commonly called environmental consciousness could be described as subterranean consciousness—the awareness that we are in a very real sense not on the earth but inside it.

When I wrote these words, in the late 1980s, I never imagined that environmental consciousness would develop so rapidly and broadly. Even less did I imagine that it would so much resemble subterranean consciousness. But this is the essence of concern about climate change: consciousness that our habitat extends to the upper reaches of the atmosphere, and that in this sense we indeed dwell below the surface of the earth.

For an academic book, *Notes on the Underground* was, as the expression goes, well received. If it had any role in stimulating public awareness, it was a bit part. (It did get rave reviews in various editions of *The Whole Earth Catalog.*) The two most decisive events in raising public awareness of climate change, however, occurred just before the book's 1990 publication. In 1988 the United Nations established an Intergovernmental Panel on Climate Change (IPCC) to track and report on research into just-emerging predictions of global warming.[1] In 1989 Bill McKibben's book *The End of Nature* was published by Random House (after being serialized in *The New Yorker*). *The End of Nature* is widely recognized as the first account of the dangers of global warming written for a general audience.

From then on, the parallel processes of scientific research and civic education have continued to shape public environmental consciousness. As I write this afterword, in the spring of 2007, the IPCC has just released a comprehensive report detailing the already evident effects of global warming—melting glaciers, rising seawater, stronger storms.[2] Citizens around the United States will soon be demonstrating in their local communities for government action on climate change in a nationwide "step it up" campaign initiated by Bill McKibben. *The End of Nature* has been translated into twenty languages. Around the world, consciousness of living "beneath the atmospheric ocean, in a closed, sealed, finite environment" has never been keener. This is the result of a still emerging partnership between science and the humanities: science, in developing understanding of how environmental systems work; humanities, in developing understanding of the implications for human life.

---

1. In the summer of 1988, James Hansen, director of NASA's Goddard Institute for Space Studies and "dubbed NASA's top climate scientist by the media," testified before the Senate that greenhouse warming was affecting the global climate (*Science* 316, 8 June 2007, p. 1412). The entire *Science* article presents an update of events involving Congress and the IPCC.

2. James Kanter and Andrew Revkin, "Scientists Detail Climate Changes, Poles to Tropics," *New York Times,* April 7, 2007. This report followed one released in early February 2007 summarizing scientific evidence for climate change. See Bill McKibben, "Warning on Warming," *New York Review of Books* 54, no. 4 (2007).

A third bedrock passage in *Notes on the Underground* underscores the need for this partnership. This passage comes at the end of the second chapter, which tells how the depths of the earth and of time were simultaneously opened to scientific investigation (p. 50). In that chapter I criticize some other scholars for tending to equate science with Baconianism—that is, with an "intrusive, aggressive, exploitative" model of inquiry. I contend that there are many varieties of science besides the Baconian model—notably what environmental historian Donald Worster[3] has called "arcadian" science in the ecological tradition—and that

The discovery of deep time has forced the human imagination to grow, not to diminish. The advancement of scientific knowledge by no means destroys the sense of awe that comes from intruding into realms of lost time. . . . The far more serious threat to nature's life comes from technological progress, at least as progress has been generally understood in modern times. . . . Our primary concern . . . should be to prevent the objective death of nature. For that we need not less science but more—at least, more science of the "arcadian" variety.

Since the publication of *Notes,* environmental science has advanced knowledge of the consequences (often unintended) of so-called technological progress through well-established processes of scientific research: accumulating evidence in peer-reviewed publications, then subjecting it to vigorous debates in the scientific community. This knowledge is essential, but not sufficient. The procedures of science are startlingly precise in locating problems of chemical pollution—poisonous substances measured in parts per billion—but they are not designed to measure and assess non-material forms of environmental degradation.[4] On the contrary, they are designed to exclude from consideration the welter of ordinary, daily, holistic experiences of the world, just as they are designed to separate matters of value from matters of fact.[5] These refusals are the source of its powers. They are also the source of its limitations.

Notes tries to show that human experience needs to be a full partner of scientific knowledge in understanding the new habitat of humanity.

---

3. In the text (p. 27), Worster's first name is incorrectly given as Daniel rather than Donald. Because of production requirements, this mistake remains. I now owe this distinguished scholar not one but two apologies.

4. See my discussion of the distinction between pollution and development as sources of environmental degradation in "Cultural Origins and Environmental Implications of Large Technological Systems," *Science in Context* 6, no. 2 (1993): 377–403.

5. Alfred North Whitehead, *Science and the Modern World* (Free Press, 1925), p. 94; David Abram, *The Spell of the Sensuous* (Vintage Books, 1997 [1996]), p. 32.

At present, the partnership is lopsided. Discussions of environmental issues routinely assume an opposition between objective and subjective factors, between nature and culture. "Nature" is understood as the given world outside humanity, the realm of objective reality, and the subject matter of scientific research. Everything else is "culture"—the world of human beings (feelings, thoughts, perceptions, habits, organizations), the realm of subjectivity, mused over by the social sciences (mimicking but inferior to the "hard" sciences) and the humanities (non-scientific, messy, sometimes soothing, but having nothing to do with serious knowledge). Over and over again, whatever the issue, this sorting takes place before the discussion really begins. Matter is real. What matters to people is apparently less so.

But nature is not outside of and separate from humanity. We live within nature. For millions of years, human beings have evolved along with the earth, and as a result we are exquisitely attuned to its conditions. Our psychological and social characteristics—human nature, so called, and human culture—are as objective and real as anything else on the planet. Although consciousness of an environmental crisis has developed at an astonishing rate since 1990, the humanistic dimensions of environmental issues remain poorly understood. The webs of relationships between human beings and the earth that involve values, memories, symbolic representations, meanings, histories, anticipations, social organizations, and cultural connections—all these critical relationships cannot be framed within the normal procedures of science, and so they are often not framed at all.[6]

The environmental crisis of our time is not just a pragmatic, an ideological, or a material one. It is also a crisis of the natural world as a cultural category. My highest hope for the new edition of *Notes* is that it will speak for the humanistic dimensions of environmentalism and for the need to reintegrate culture and nature, subjectivity and objectivity, in a disturbed and dangerous world.

This brings us to a fourth defining passage, found at the end of chapter 5, which analyzes degeneration and defiance in nineteenth-century subterranean tales. These pages recount an episode near the end of E. M. Forster's 1909 short story "The Machine Stops" in which a vicious "mending device," launched by the underworld's tyrannical Committee of the Machine, destroys a gentle valley in Wessex. Commenting on this episode, I wrote (pp. 146–148):

---

6. Abram, passim.

What is unique about the natural environment, what can never be replaced by the technological one, is its independence of the social order. . . . Natural despoliation is not just a result of economic pressures; it is also a political action aimed at removing a source of subversion. . . . [Natural destruction may be recognized as] a requirement, rather than an accidental byproduct, of a social and political system that excludes from its moral frame of reference values that are not material, utilitarian, and presentist.

I remember getting worked up as I wrote these paragraphs, which is probably why the language is so convoluted. I was aware that I was repeating a centuries-old protest, the Romantic protest against the exclusion of value, defined as "the intrinsic reality of an event, . . . from the essence of matter of fact."[7] I was also aware that this protest has never been only a conceptual one, but that it has also led to repeated acts of resistance against the powers—military, political, economic—that deprive people of their familiar, productive place in the world by uprooting them or destroying their habitat or both.

Recognizing just how deep and widespread this protest is, far from calming my anger, made it stronger. Mourning for a ruined landscape, or lamenting the loss of one's home, is routinely disregarded just because it is familiar. Such expressions of sorrow over habitat loss are labeled regressive, nostalgic, and sentimental whining—or, worse, reactionary appeals to nationalistic or ethnic identity or class-based selfishness, as if the long-evolving connection between human and earth were an effete bourgeois luxury.

In *Notes* I describe such fatalistic submission to environmental destruction as the triumph of the "social sublime." Because it combines aesthetic and political experiences, the concept of social sublimity is useful to describe events orchestrated by powerful humans to convey the appearance of more-than-human power, thereby instilling feelings of shock and awe in the relatively powerless. For example, Forster's short story "The Machine Stops" shows how feelings of helplessness are aroused and reinforced through images and experiences of individual dependence on The Machine. In our own time, the process of "technological change" has become a vast display of social sublimity. Just to invoke "change" seems to justify its fatal inevitability—as if it were all one huge process, as if it justified every project of destruction, as if it came out of nowhere.

---

7. Whitehead, *Science and the Modern World,* pp. 93, 94.

The eighteen years since the publication of *Notes* have seen significant development of environmental consciousness. Those years have been much less productive of effective environmental politics, at least in the United States. The process of revisiting *Notes* has sharpened my awareness of some underlying, persistent sources of confusion in current environmental politics. I did not write *Notes* as an environmental activist, nor would I describe myself as one now. *Notes* is an academic study—but the questions it raises are not only academic.

As I have already remarked, a primary source of confusion in environmentalism is the difficulty of articulating the significance or even the reality of the whole range of values, beyond the dominant material ones, that human beings seek and find in their earthly home. Another, also mentioned already, is the veil of confusion thrown over events when they are all described (in the language of social sublimity) as inevitable change. It is true that change is a constant. Human beings—along with other creatures—have been altering the conditions of their lives for millions of years. But it is also true that recent changes in the earth caused by human actions are unprecedented in scope, scale, pace, and impact. The general inevitability of change cannot be used to justify all particular changes, especially if they threaten the habitat of humanity.

Finally, there is a confusing lack of realism about the role of power in history. American environmental politics places undue emphasis on consumer choice—as if environmental despoliation were only a result of lifestyle, as if consumers always have genuine choices, as if there were not social structures organized around powerful economic, commercial, and military priorities. Consumer-oriented activism is necessary, but it is inadequate and can even be diversionary if it avoids confronting power structures. These have consequences as real and predictable as anything in the physical universe.

In combination, these confusions cripple effective environmental politics. If a wide range of human experiences are dismissed as sentimental, nostalgic, and even reactionary; if resistance to those events is deemed futile because change is a constant, just as complaining about change is a constant; if the reality of power in human affairs is evaded; then the ground of politics is sapped by ambivalence.

On the last page of *Notes*, in the fifth and final bedrock passage, I write:

Our environment will inevitably become less natural; the question is whether it will also become less human.

The goal of environmental politics is not saving the planet, but creating (and constantly recreating) a human world. The environmental crisis of our time is not only about science, technology, and economics; it is also about understanding human beings and human organizations. This is the multidisciplinary challenge of the century, and the humanities are a crucial part of it.

Technology, Environment, and History

Chapter 3 of *Notes* describes late-nineteenth-century Western Europe as a transformative time and place in the emergence of a new habitat for humanity. It describes multiple forms of excavation related to mining, industry, sewerage, transport, and other large systems, all of which created a new infrastructure, at once technological and social, of modern life. The chapter argues that these activities were so intensive and impressive that they encouraged some thoughtful observers to conclude that humanity had reached a turning point in history, the point where a predominantly natural world (in the conventional sense of non-human nature) was giving way to a predominantly built one. For them, imagining humanity's descent into a subterranean world seemed an appropriate analogy for this historical point of no return.

Notes was published at the time when other historians of technology were starting to give more attention to technology as the built world, the built environment, or the human-built world (to use three of the more common terms, all of which have the limitation of identifying the crucial term 'technology' with its material consequences). The concept of "technology as environment" dates from the 1950s, but for several decades it tended to be limited to urban infrastructure.[8] In the 1980s, however, historians of technology began to emphasize systems, rather than machines or devices, as the primary unit of analysis. Systems might have their nodes in cities, but they extended well beyond. For example, Thomas P. Hughes showed in his pathbreaking book on electrical networks how these "networks of power" connect cities with regions and nations.

8. William F. Ogburn, "Technology as Environment," *Sociology and Social Research* 41 (1956): 3–9. For a summary of some of those studies at the time *Notes* was published, see Eugene P. Moehring, "The Networked City: A Euro-American View," *Journal of Urban History* 17, no. 1 (1990): 88–97. Also published in 1990 was *From Artifact to Habitat,* ed. G. Ormiston (Lehigh University Press and Associated University Presses).

In the later 1980s and the 1990s, other historians began to consider the world as made up of multiple, interactive layers of technological systems, and to publish articles and books on the convergence of technology and the environment.[9] Hughes himself encouraged this conceptual evolution by defining technological systems in an expansive and adventurous way. He also increasingly emphasized the cultural and artistic dimensions of engineering. His work in these directions culminated in the 2004 publication of his book *Human-Built World: How to Think about Technology and Culture* (University of Chicago Press).

I was fortunate to follow these developments from a supportive niche at the Massachusetts Institute of Technology. In the early 1980s I enjoyed a research fellowship there in the Program in Science, Technology, and Society. I met David Nye, who would go on in the 1990s to write and edit an impressive array of books exploring "the technological sublime," "technologies of landscape," and "second creation."[10] When I joined the Writing Program's faculty in the later 1980s, Tom Hughes, a regular visitor to MIT, helped provide intellectual leadership for a multi-year workshop on technological systems organized by Joel Moses (then Dean of Engineering), which I was invited to join. I also attended a lively multi-year seminar on humanistic environmentalism organized by STS colleagues Leo Marx, Jill Ker Conway, and Kenneth Keniston.[11] Other MIT colleagues, including Deborah Fitzgerald in STS and Harriet Ritvo in History, were doing exciting work in agricultural, environmental, and rural history. Eventually they too organized a vibrant and still-ongoing seminar on these topics.[12]

9. For example, Tom F. Peters, *Building the Nineteenth Century* (MIT Press, 1996); Helen Meller, *European Cities 1890–1930s* (Wiley, 2001). For a useful review article, see Jeffrey Stine and Joel Tarr, "At the Intersection of Histories: Technology and the Environment," *Technology and Culture* 39, no. 4 (1998): 601–640. Stine and Tarr also edited a special issue of *Environmental History Review* (18, no. 1, 1994) on "Technology, Pollution, and the Environment." Another review essay with a more limited compass is Brian Black, "Construction Sites: Environment, Region, and Technology in Historical Stories," *Technology and Culture* 40, no. 2 (1999): 375–387.

10. David E. Nye, *American Technological Sublime* (MIT Press, 1996); *Technologies of Landscape* (University of Massachusetts Press, 1999); *America as Second Creation* (MIT Press, 2003).

11. The series was supported by funds from the Mellon Foundation and led to the publication of a collection of papers from the seminars: *Earth, Air, Fire, Water*, ed. J. Conway, K. Keniston, and L. Marx (University of Massachusetts Press, 2000).

12. The series has been supported by funds from the Sawyer Foundation. For a review of the history of American agricultural technology, see Deborah Fitzgerald, "Beyond

While historians of technology were studying technology-as-environment, environmental historians were studying environment-as-technology. During the 1990s the history of technology and environmental history both grew taller and closer, like neighboring trees whose canopies begin to connect. Though the roots of environmental history are considerably older,[13] a new and flourishing phase of growth began in the late 1970s along with the environmental movement in American politics. Much of this scholarly activity involved American historians looking at the American landscape, especially the American West. In the late 1980s and the early 1990s, environmental history began to shed its disproportionate attachment to the American experience, and along with it a strong ideological attachment to supposedly unspoiled nature and wilderness.[14]

The latter trend was encouraged, not without controversy, by scholars such as William Cronon, and Richard White. Cronon's 1995 essay "The Trouble with Wilderness,"[15] published not that long after McKibben's book *The End of Nature,* makes a similar point about the need to overcome mindsets and behaviors that unrealistically separate nature and culture. Cronon's analysis aroused considerable protest, especially from environmentalists who felt he was devaluing "wilderness." But it also proved increasingly persuasive as a description of the world we live in. Richard White's 1999 Tanner Lecture "The Problem with Purity" takes some jabs at Cronon but ultimately echoes his message that historians must accept "a mixed and

---

Tractors: The History of Technology in American Agriculture," *Technology and Culture* 32, no. 1 (1991): 114–126.

13. Pamela O. Long, "The Annales and the History of Technology: *Annales d'histoire économique et sociale* 7 (November 1935), Les techniques, l'histoire et la vie," *Technology and Culture* 46, no. 1 (2005): 177–186.

14. The journal *Environmental Review* began publication in 1976. In 1989 it was renamed *Environmental History Review.* Eventually it merged with *Forest and Conservation History* to become the current international journal *Environmental History.* The American Society for Environmental History was founded in 1982. In 1988 Donald Worster edited an influential collection, titled *The Ends of the Earth* (Cambridge University Press), which showed the wide range of topics that were possible in the field; Worster's own contribution to that volume ("Doing Environmental History") set out a program for research. See Alfred W. Crosby's critique of the "American tilt" and national parochialism of much environmental history (pp. 227–228) in "Afterword: Environmental History, Past, Present, and Future," in *City, Country, Empire,* ed. J. Diefendorf and K. Dorsey (University of Pittsburgh Press, 2005).

15. William Cronon, "The Trouble with Wilderness; or, Getting Back to the Wrong Nature," in *Uncommon Ground,* ed. W. Cronon (Norton, 1995).

dirty world in which what is cultural and what is natural becomes less and less clear and . . . hybrids of the two become more and more common."[16]

An early sign of the convergence of technological and environmental history was the 1991 annual meeting of the American Society for Environmental History, organized around the theme of "The Environment and the Mechanized World." That meeting brought together urban historians, agricultural historians, historians of science, students of environmental politics, and political theorists, among others, on a wide range of topics: "applied environmental history," dams, nuclear power, imaging the earth from space, sewers, wildlife management, and African-American environmental activism in Gary, Indiana.[17] A more recent indicator is the formation in 2000 of a special-interest group dubbed "Envirotech" under the auspices of the Society for the History of Technology.[18] All the while, other disciplines and subdisciplines have been contributing to studies of the hybrid human lifeworld. The discipline of geography has been revitalized and fruitfully connected with historical studies.[19] Landscape studies have engaged literary historians, art historians, cultural geographers, and social scientists.[20] If

16. My source for White's comments is http://www.tannerlectures.utah.edu. McKibben is one of the environmentalists uneasy with the implications of Cronon's questions about the meaning of wilderness. See Bill McKibben, *Wandering Home* (Crown, 2005).

17. For recent summaries of environmental history, see Jeffry M. Diefendorf and Kurk Dorsey, "Challenges for Environmental History," in *City, Country, Empire,* ed. Diefendorf and Dorsey. For more recent reviews of environmental history, see J. R. McNeill, "Observations on the Nature and Culture of Environmental History," *History and Theory* 42, no. 4 (2003): 5–43; Ted Steinberg, "Down to Earth: Nature, Agency, and Power in History," *American Historical Review* 107, no 3 (2002): 798–820; Carolyn Merchant, *The Columbia Guide to American Environmental History* (Columbia University Press, 2002); and the 29 brief essays published under the heading "What's Next for Environmental History" in *Environmental History* (10, no 1, 2005).

18. A current book project being sponsored by Envirotech is titled "Ties that Bind: Environment and Technology in History" [see Dorotea Gucciardo and John Crosmun, "Conference Report: Taking Stock of the *Longue Durée,*" *Technology and Culture* 48, no 1 (2007): 153–157]. For recent reviews of history of technology as environment, see Theodore R. Schatzski, "Nature and Technology in History," *History and Theory* 42, no. 4 (2003): 82–93. See also comments on technology in J. Donald Hughes, *What Is Environmental History?* (Polity, 2006), pp 48–53.

19. For an excellent review with an extensive bibliography, see Alan R. H. Baker, *Geography and History* (Cambridge University Press, 2003). See also H. C. Darby, *The Relations of History and Geography* (University of Exeter Press, 2002).

20. Some reviews can be found in Thomas Greider and Lorraine Garkovich, "Landscapes: The Social Construction of Nature and the Environment," *Rural Sociology* 59 no. 1 (1994): 1–24; William Norton, *Explorations in the Understanding of Landscape* (Greenwood, 1989); Ian D. Whyte, *Landscape and History since 1500* (Reaktion Books, 2002).

there is ever an example of the value of multidisciplinary research, it is at the intersection of technology and science studies, environmental history, geography, landscape studies, and the arts, as they combine to form a cable of inquiry into the origins and implications of the new human habitat.

Today, this cable of inquiry is much thicker and stronger than when I wrote *Notes on the Underground*. At the time *Notes* was published, only Wendy Lesser had written in a sustained way about connections between subterranean experiences in literature and in history.[21] Now, 18 years later, David Pike has written three books further exploring these connections, two of which devote particular attention to subterranean Paris and London in the nineteenth century.[22] Donald Reid's 1991 book *Paris Sewers and Sewermen: Realities and Representations* (Harvard University Press) won the "outstanding scholarly book" award from the Society for the History of Technology in 1992.

Larger claims of an unprecedented environmental transformation beginning around the year 1900 have been documented in J. R. McNeill's book *Something New under the Sun: An Environmental History of the Twentieth-Century World* (Norton, 2000), which surveys this transformation in its multiple manifestations: industry, agriculture, energy use, settlement patterns, demographics, soil, air, water, non-human flora and fauna, and other resources. According to Jeffry Diefendorf and Kurk Dorsey, the success of McNeill's "ambitious project" was possible "because he had a vast body of scholarly literature from around the world upon which to draw.... It seems likely that such a comprehensive study would not have been possible much earlier."[23]

Taken together, the recent research in technological and environmental history supports a fundamental claim of *Notes*: that there has been a recent and profound change in the human relationship with the earth. The labels are familiar: second industrial revolution, second agricultural revolution, globalization, urbanization. What is new is the understanding of how these processes simultaneously and interactively transformed the world.[24]

This is how research works, in the humanities as well as in the sciences: through the gradual and collective accumulation of knowledge, and

---

21. Wendy Lesser, *The Life below the Ground* (Faber and Faber, 1987).

22. David L. Pike, *Subterranean Cities* (Cornell University Press, 2005); *Metropolis on the Styx* (Cornell University Press, 2007).

23. Diefendorf and Dorsey, p. 5. See especially the summary of global measures of change on pp 360–361.

24. For a discussion, see Rosalind Williams, *Retooling* (MIT Press, 2002), pp. 20–24.

---

through the leadership of scholars who do more than most to interpret the meaning of this knowledge, to underscore its implications, and to provide intellectual direction. During the later years of the twentieth century, the history of the built world was itself built through work that is a model of multidisciplinary scholarship.

Literature and History

At the time I was writing *Notes,* I was teaching technical, essay, and expository writing to MIT undergraduates as a faculty member of the Writing Program (since renamed the Program in Writing and Humanistic Studies). Not until ten years later did I join the STS faculty and begin to teach history to both undergraduate and graduate students. Much of this second phase of my teaching life has involved co-teaching with my STS faculty colleague Leo Marx, whose seminal book *The Machine in the Garden* is briefly (too briefly) mentioned on pp. 18 and 19 of *Notes* in a discussion of differences in environmental imagination between the New World and the Old.

During the past five years, Leo and I have regularly taught topics in history and literature to graduate students from several departments at MIT, and also to students from Harvard. It is often asserted that the interplay of teaching and research is the special strength of higher education in research universities. This is true not only for the students but also (and even more) for the faculty. In advancing my understanding of the connections between history and literature, nothing has been more important since the publication of *Notes* than teaching with Leo.

Teaching is what the formal arrangement is called. In reality, everyone in the class is exploring questions for which none of us knows the answers. The basic questions are two: "Why has the word and concept 'technology' risen to such dominance in contemporary discourse?" and "What is the relationship between history and literature?"[25] One exercise that has proved especially useful in addressing both questions is to assign William Cronon's *Nature's Metropolis* (an environmental history of Chicago and its hinterland in the late nineteenth century) in conjunction with Theodore

---

25. On the former question, see Leo Marx, "Technology: The Emergence of a Hazardous Concept," *Social Research* 64, no. 3 (1997): 965–988. On the latter, see Leo Marx, afterword to second edition of *The Machine in the Garden* (Oxford University Press, 1989 [1964], pp. 367–385.

Dreiser's *Sister Carrie* (a novel set in the same time and place). We ask "What do we learn about the history of Chicago and the Midwest from Cronon that we don't learn from Dreiser, and vice versa?" Such comparisons invite reflection on the relationships between social structures and individual consciousness; on values, motivations, and constraints built into the world and then reinforced by what has been built; and on the indispensable role of literature in showing the integrated effects of many simultaneous changes, as experienced by human beings who use their perceptions to master the world while only dimly aware of its powerful effects on them.[26] A novel probes these experiences with a very short wavelength. Dreiser's reveals, in emotional depth and complexity and fine detail, how Sister Carrie and other characters confront the changing environment of Chicago as lived experience.

The literature featured in *Notes* provides a different kind of revelation, because it is explicitly fantastic. If all literature tends toward allegory, using concrete particulars to convey generalizations, the subterranean tales of *Notes* are overtly and strongly allegorical. They do not probe the familiar "real" world. Instead, they construct a parallel world in order to explore possibilities, probabilities, and projections of the familiar one. As quoted in the first chapter of *Notes* (p. 20), the anthropologist Clifford Geertz describes this procedure as "neither more nor less than constructing an image of the environment, running the model faster than the environment, and then predicting that the environment will behave as the model does."

This explanation too is an analogy—one that takes us back to the relationship between science and the humanities. Humanists are shy, for good reason, of claiming anything like scientific authority for their scholarship, when standards of evidence and validity are in so many ways incommensurate. However, environmental science too depends largely on predictive models—on "images of the environment," in which human and non-human factors are both essential elements. There is no laboratory apparatus, short of the planet or indeed the universe, on which to run an experiment separating human-generated from other causes of climate change. Historians such as J. R. McNeill and earth scientists alike deal

26. Here I am paraphrasing Donald Lowe. He writes of the bourgeois who "approaches the world from the inside, using the available content of perception to comprehend and master the world, often with minimal awareness of the determination by the perceptual field." Donald M. Lowe, *History of Bourgeois Perception* (University of Chicago Press, 1982), p 18.

with evidence where human population and technologies have reached such a scale that they are intertwined with natural systems. No one can pull them apart: historian and scientist alike must integrate many different data, and they both have to make predictions based on models of current hybrid events.

The narratives featured in *Notes* emphasize social and psychological rather than physical effects of environmental change, but they do make predictions (summarized on pages 211–212)—predictions that human needs for security, authority, reality, and transcendence will all be challenged under conditions of a largely self-made habitat. But the prophetic value of these fantasies lies not only in their content, but also in their tone. It is striking that their writers, arguably among the most privileged people in the world of their day (male, white, educated, bourgeois), are collectively so anxious and melancholy. In their narratives they present evidence of material progress, as conventionally defined in the bourgeois West, but they do not celebrate it. Even such an apparently progressive writer as Jules Verne dwells on images of failure and catastrophe. Gabriel Tarde's supposed utopia is undercut by irony and ends with the tragicomic suicides of lovers who die together on mountaintops. At the height of British power, E. M. Forster imagines human disconnection from the planet so profound that he writes an elegy for the earth.

When a pattern of life is radically disrupted, "we should find counterparts to grief and mourning—a similar struggle to repair the essential thread of continuity."[27] An acceleration of change—even if it is named "progress"—also accelerates loss. The writers of these underground fantasies lived in a world of environmental transformations unprecedented in scope, scale, and pace. When their entire habitat is disrupted, their pervasive sense of bereavement is entirely predictable.

Conclusion: The Essay Form

Notes on the Underground is subtitled "an essay on technology, society, and the imagination." As I have mentioned, when I wrote it I was earning my keep at MIT by teaching writing, and so I had the essay genre on my mind from daily experience. In writing *Notes*, I deliberately opted for what one reviewer generously if somewhat backhandedly described as "the seeming

---

27. Peter Marris, *Loss and Change* (Pantheon Books, 1974), p 3.

disjunctions of the extended essay rather than the tightly ordered logicalities of the empirical monograph."[28] I would prefer to say that I was opting for flexibility and expansiveness, rather than disjunctions.

Most of the time, I believe, the essay form served well. The only part of the book where disjunctions seem to have gotten out of hand is near the end, where in a few pages (204–210) I sketch some then-current subterranean projects or proposals. Some of them have come to fruition; others have not. In either case, the section seems dated. Here it is not the essay form that is at fault; rather, I am at fault for giving in to the temptation to relate the book to current events in too direct a way. So I will now resist the temptation to recap more recent technological and cultural projects featuring subterranean sites, other than noting that I am always on the lookout for them, and that I have enjoyed writing about Boston's Big Dig and about the *Matrix* movies.[29]

Another advantage of the essay, besides its flexibility, is its fusion of objectivity and subjectivity. It can present empirical evidence, but it also encourages a more personal, individual voice for the writer than does the scholarly monograph. Essay readers need to keep their ears as well as their eyes open, to listen to the points where the tone becomes urgent and the emotional temperature begins to rise. This is especially true when one is reading an essay that one has written. I have already mentioned a passage in *Notes* where I got worked up commenting on Foster's short story "The Machine Stops." In another passage (p. 147), I describe Forster's imagined death of nature:

Nature is destined to become a ruin, a part of ancient history, a dead empire of which we now see the half-buried relics. Like decaying bodies on the gibbet, rotting rivers and ruined meadows are cautionary lessons, the emblems of superior power. Nothing is sacred but that power.

Therefore, mourning for the forgotten or ruined landscape is not necessarily a sentimental and outdated emotion.

If I had been paying attention to my own overwrought tone here, I would have paid more attention throughout *Notes* to the most obvious meaning of the underworld: it is the place of burial, the actual and symbolic realm

---

28. Bill Luckin, "Sites, Cities, and Technologies" (Review Essay), *Journal of Urban History* 17, no. 4 (1991), p. 430.

29. See Rosalind Williams, "The Big Dig," *Technology and Culture* 47, no. 3 (2006): 707–711. On the *Matrix* series, see Rosalind Williams, "Presidential Address: Opening the Big Box," *Technology and Culture* 48, no. 1 (2007): 104–116.

of the dead. How could I write a book about the underground and not say more about death?

One answer is that I was years younger, less prone to think about my own mortality, and less experienced with that of others close to me. Still, the strength of my unconscious resistance is striking. I am not alone in this. Lewis Mumford refers to the earth's surface as "the environment of life" but does not summon the courage to call the mine the environment of death. Instead he shifts the terms of the contrast and calls the mine "the environment alone of ores, minerals, metals" (quoted on p. 5 of *Notes*).

It is little comfort to recognize that I share a widespread cultural evasion. To be sure, there is quite a bit of discussion in *Notes* about fascination with death-dealing catastrophes. This is a theme of the last chapter, which explains how nostalgia for a lost organic earth can transmute into "nostalgia for disaster," as a means of restoring a supposedly more natural world, by ridding humanity of its self-generated one.[30] But this in a way is a modestly hopeful theme: imagining that a more human life can be recovered through disaster. There is another possibility, which is that the human world will perish along with much of non-human life. One critic who noticed the silence in *Notes* about death concluded his review on a more ominous note:

The world below may be our (scarcely inexhaustible) mine of wealth, and the place to which we scurry in the hope of shelter, but it is also where we bury whatever we most want to keep out of sight: radiation, unutterable destruction, our dress rehearsals for universal death.[31]

Is the quest for a self-created human habitat a death cult through which human beings seek to defy organic limitations through technological prowess? Since death is one limit we cannot master, does our anger over this limitation drive us to dig a self-constructed tomb rather than accept that of the earth?[32]

In *Notes on the Underground* "there's death on every page."[33] Every page about the underworld is a reminder not only of the death of humans,

---

30. The phrase "nostalgia for disaster" is George Steiner's, quoted and analyzed on p. 188 of *Notes*.

31. Patrick Parrinder, "Troglodytes," *London Review of Books,* 25 October 1990, p. 24.

32. George Kateb, "Technology and Philosophy," *Social Research* 64, no. 3 (1997): 1225–1246.

33. This is a reference to a 1989 song by the rock group Queen, "My Life Has Been Saved."

but also, by their absence, of the implied death of creatures with whom we share the earth's surface. They are hardly ever mentioned as dwellers in either real and imaginary underworlds. The miserable existence of ponies and the ominous role of canaries in coal mines only underscore the point. Almost all non-human life forms so familiar to us on the surface of the earth are eerily absent from its silent depths.

Notes thereby raises a fundamental dilemma inherent in the current environmental crisis. The connection between environmental sustainability and human population growth has often been observed. If twentieth-century rates of population growth had prevailed since the beginning of agriculture, "the earth would now be encased in a squiggling mass of human flesh, thousands of light-years in diameter, expanding outward with a radial velocity many times greater than the speed of light."[34] One cause of overpopulation may be a high birth rate. Another cause may be a low death rate.

The deepest human desire—for life itself—makes us want to prolong existence for ourselves and for those we love. As a society, we pour resources into ways of evading or delaying mortality. But the more human beings multiply and fill the earth, the more its resources are consumed and other species are crowded out. The health of individuals and that of the planet cannot both be sustained indefinitely. Environmental sustainability requires that humans regularly die.

Now that I am somewhat more willing to confront this dilemma, I realize that *Notes on the Underground* is less about a particular space than about a particular time. It is about an historical event that does not yet have a name, an event for which existing terms are inadequate. There is a "semantic void" in our language similar to that which gave rise to the current reliance on the word 'technology' to point to a totalizing array of experiences and agencies that cannot be named by more limited, physically concrete words, such as 'machine' or even 'industry'.[35]

For now, the best I can do in giving a name to this event is to call it the establishment of the human empire on earth. Human beings have always been altering the face of the earth. Now we are changing the fate

---

34. McNeill, p. 9, referring to Carlo Cipolla, *The Economic History of World Population* (Penguin Books, 1978), p. 89.

35. This is a central argument of Leo Marx's "Technology: The Emergence of a Hazardous Concept."

of the earth.[36] The ability to do this, and the extent of human dominance, are unprecedented. There are many physical measures of this ability and dominance,[37] but their defining characteristic is nothing material at all. This event is ultimately one of consciousness: the realization, among human beings, that our needs, desires, works, and actions will henceforth rule the fate of everything on the planet, including our own.

Notes goes as far as it can in using the spatial analogy of the underworld to understand the meaning of this event. There is much more to be said about it, but this will require fresh sources of evidence and insight. And that is another book.

---

36. See I. G. Simmons, *Changing the Face of the Earth* (Blackwell, 1996 [1989]) and (especially) the three-volume *Man's Role in Changing the Face of the Earth*, ed. W. Thomas Jr. (University of Chicago Press, 1956). See also Jonathan Schell, *The Fate of the Earth* (Knopf, 1982).

37. McNeill's book *Something New under the Sun* presents many such measures. On other ways of measuring "the human domination of our planet," see Peter Kareiva, Sean Watts, Robert McDonald, and Tim Boucher, "Domesticated Nature: Shaping Landscapes and Ecosystems for Human Welfare," *Science* 316 (29 June 2007): 1866–1869.

Index

Narrative analysis and narrative structures, 19–21, 153, 183
Naturalistic fiction, 161–163
Natural scenery, 114, 156, 185, 201, 203
in "The Machine Stops," 135, 145–146, 148
and sublimity, 84–85, 98, 154, 187
Nature, 107, 110–111, 114, 140, 183–184, 202–203, 206
banishment of from closed interiors, 91–92, 94, 97–98, 103, 109, 153
conquest of, 125–126, 132, 142, 147, 169, 174, 200–201
death of, 198
domination of in Verne, 100–101, 166, 171, 194, 197
female images of, 24–25
journeys back to, 144–145, 182–185
as moral authority in Forster, 132–133, 135, 145–150, 212
purification of in Tarde, 135, 145–146, 148
return to through disaster, 202–204
as source of hazard, 189–192, 195, 197–198
Neanderthal skull, 36–38

Orpheus, 8
Orwell, George
1984, 175, 202
The Road to Wigan Pier, 69–70, 139–141

Paleontology, 23, 29–30. See also Cuvier
Paris, 159, 201
Baudelaire on, 109–110
sewers of, 47, 70–72, 82, 159, 199, 204
subways of, 73–75, 79
Pastoral and pastoralism, 3, 18–19, 115, 122, 182, 184
in Forster, 146
revolutionary, 183
in Tarde, 137
utopian, 124
in Verne, 166, 168

Pellucidar. See Burroughs
Petrie, William Flinders, 40, 42
Piranesi, Giovanni Battista, 92–94, 177
Pitt-Rivers, Augustus Henry, 42
Plotlines, structural congruence of, 16, 19, 153
Poe, Edgar Allan, 45, 94, 121, 188
and Symmes, 13–14
Verne on, 14–15
Political fiction, 164–165, 173
Politics, 65, 141–145, 164. See also Anarchism; Socialism; Revolution; Terrorism
of Bulwer-Lytton, 129–132
environmental, 146–148, 203
and imagination of disaster, 189, 200, 203–204
middle class and, 65–66, 153
and personal consciousness, 182–185
of Verne, 169–172
in works of Wells, 172–173, 177–182
Population, 130, 133, 137–139, 143–144, 196, 200–202, 206
Post-structuralists, 49
Prestwich, Joseph, 36
Progress, 49–50, 64–66, 79–80
and apocalyptic tradition, 188–189, 201–204
in The Coming Race, 127, 129, 131
and destruction, 63–64, 90, 123, 125, 193, 196, 203
excavation as emblem of, 53–54, 204
Forster's definition of, 132
Orwell on, 140–142
in second industrial revolution, 70, 108
tunnels as examples of, 58, 60
in Underground Man, 136, 195
in Verne's works, 166, 172
Proserpine, 8

Quirk, John J., 107–108

Realistic literature, 177. See also Industrial novels

Printed in the United States
by Baker & Taylor Publisher Services